HOW TO USE THIS BOOK

- This book contains thousands of carefully constructed questions.
- The questions are organised into curriculum-aligned Question Sets that deal with individual learning objectives.
- The Question Sets are structured and categorised by topic and subtopic to help you plan using Learning by Questions.
- Each Question Set is set out in this book with all its questions, answers and diagrams or illustrations to allow you to check suitability for your class.
- Once you have found the perfect Question Set you'll be able to find it on our website at www.lbq.org using the Quick Search References.

Everything in this book is online at www.lbq.org

LbQ can be delivered in these ways:

Self-Paced Questioning

Set a task for your pupils – they each receive questions on their devices, respond at their own pace and receive instant feedback.

Ad Hoc Questioning

Pose questions on the spur of the moment – your pupils respond on their devices.

Teach

Turn any question into a slide, ideal for modelling concepts on your classroom display.

CW01663523

When your pupils hit a challenging question,
you'll know about it.

Drill down to see every answer and be on top of every misconception.

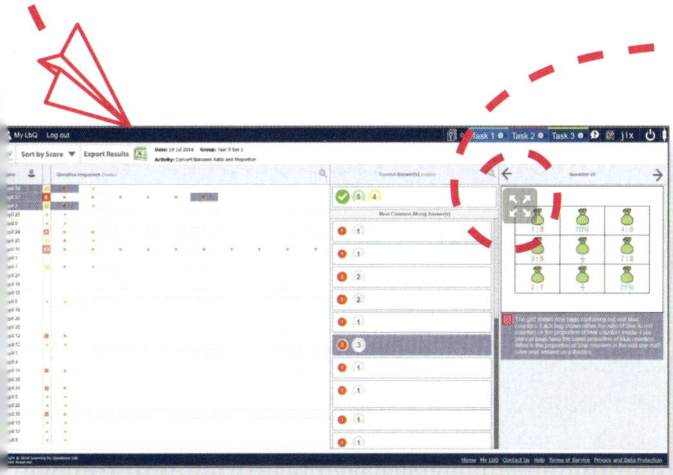

Pause, intervene, explore, explain and model using 'Teach' mode.

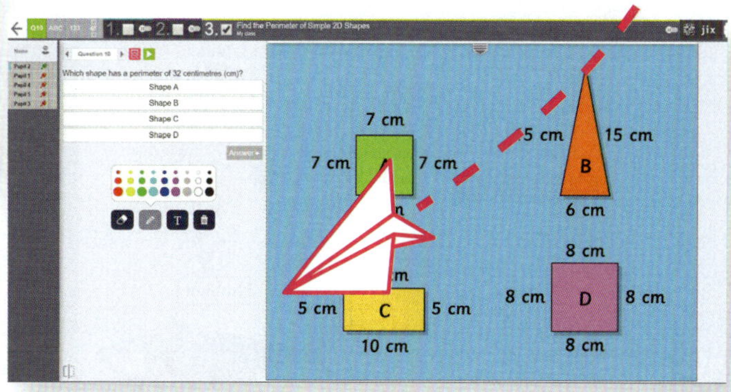

Try the question again with Ad Hoc questioning.

Everything in this book is online at www.lbq.org

SE THE BOOK TO PICK THE RIGHT TASKS FOR YOUR CLASS AND
READY FOR INTERVENTION OPPORTUNITIES.

Vant to build on an existing lbq.org question?

\nnotate or add
ext to explain,
\nodel, modify and
xtend...

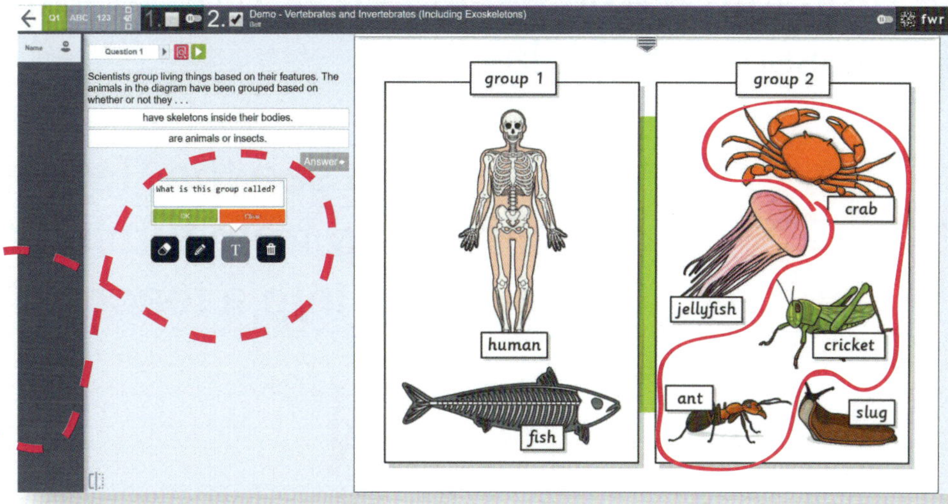

...and use Ad Hoc to send as a new question to your class.

Make your own questions on the fly.

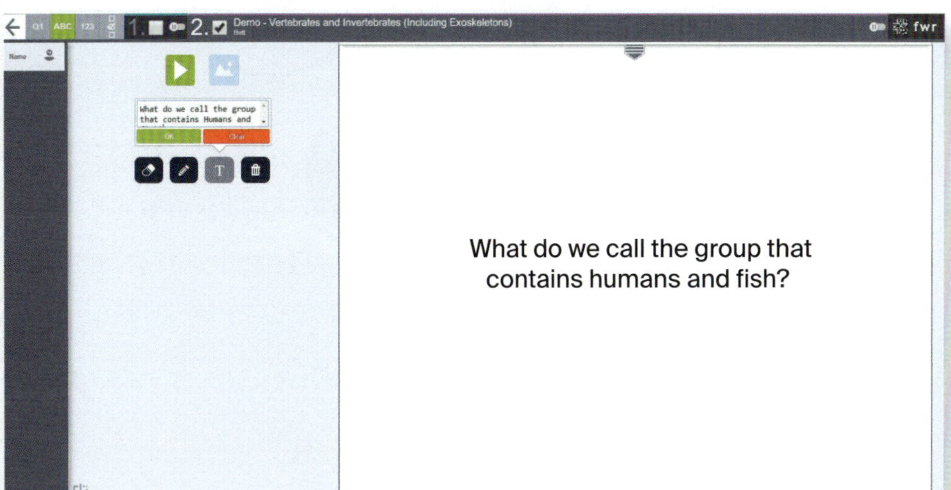

Use our Teach Tools to create the right question at the right time.

Write a question, draw a question or just ask a question.

Forget **'hands up'.** With Ad Hoc questioning everyone answers, every question, every time!

SE THE BOOK TO CHOOSE, ORGANISE AND REHEARSE
JESTIONS TO TEACH IN YOUR CLASSES.

Teach

If you don't have pupil devices, you can still turn any question into a whole-class teaching resource on your classroom display.

Wherever you see the 'Teach' icon, you're one click away from turning a question or Question Set into a teaching resource.

Each question can be an ideal teaching point.

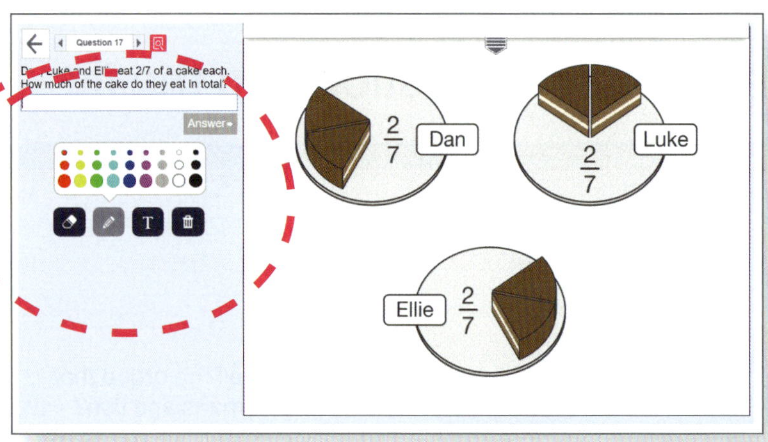

Annotate to explain, model and work as an example.

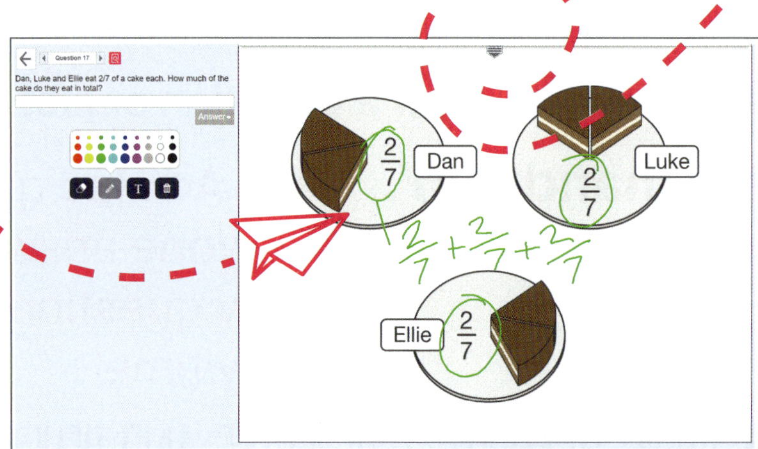

Use the pull-down pad to construct your own questions.

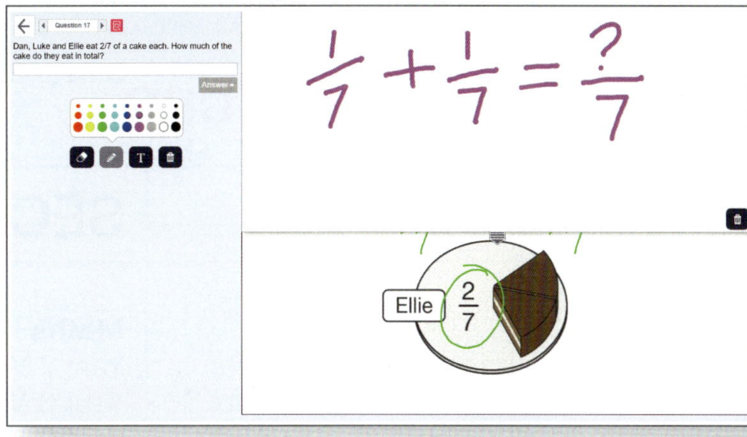

Work through multiple questions like a slide show.

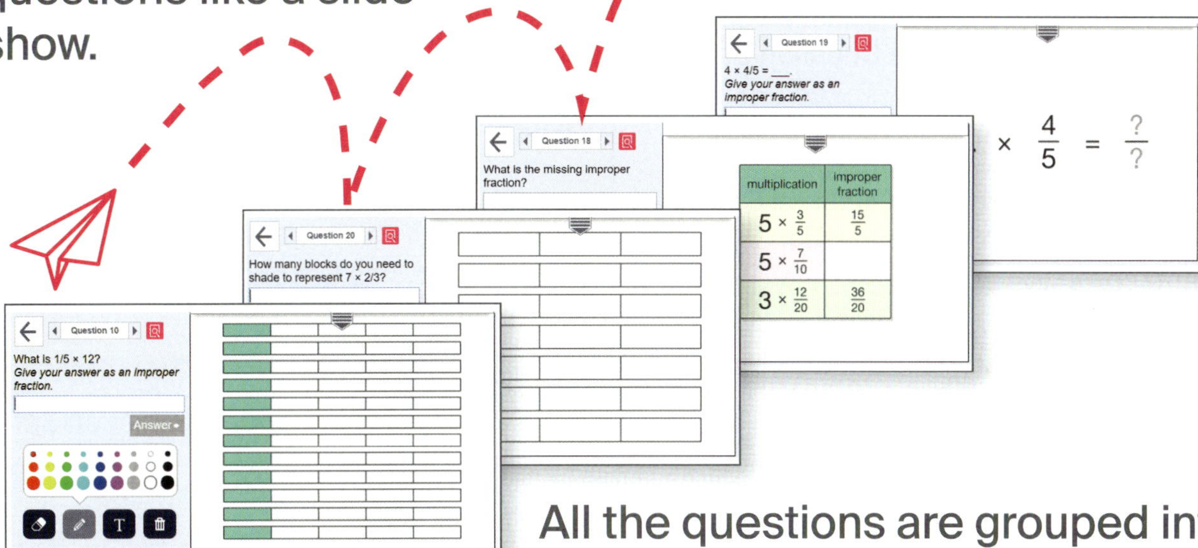

All the questions are grouped into carefully scaffolded Question Sets and provide great support for progression from lower-order to higher-order thinking skills.

TEACH MODE IS **FREE** TO ALL REGISTERED USERS

Register FREE at lbq.org

SE THE BOOK TO HELP FIND, ORGANISE AND REHEARSE UESTIONS TO INCLUDE IN YOUR LESSONS.

Other Titles in the Series

Learning by Questions

PRIMARY KS2

Maths
Year 3 Mathematics Primary Question Sets
Year 4 Mathematics Primary Question Sets
Year 5 Mathematics Primary Question Sets
Year 6 Mathematics Primary Question Sets

Science
Years 3&4 Primary Science Question Sets
Years 5&6 Primary Science Question Sets

English
Years 3&4 English Primary Question Sets
Years 5&6 English Primary Question Sets

SECONDARY KS3

Maths
Year 7 Mathematics Secondary Question Sets
Year 8 Mathematics Secondary Question Sets
Year 9 Mathematics Secondary Question Sets

Biology
Years 7–9 Biology Secondary Question Sets

Chemistry
Years 7–9 Chemistry Secondary Question Sets

Physics
Years 7–9 Physics Secondary Question Sets

English
Years 7–9 English Secondary Question Sets

Quick Search Reference Guide
All Question Sets, all subjects, all years

KS1 & KS4 Question Sets can also be found on our website
www.lbq.org

See www.lbq.org/books for title availability

Tailored Instant Feedback with Every Question

Feedback corrects misconceptions and reinforces learning to accelerate progress.

LbQ's digital platform provides pupils and teachers with instant feedback.

Our feedback is designed to reflect the kind of specific feedback a teacher would provide, given the time to do so.

Our feedback gives specific guidance and pupils are given the opportunity to retry the question.

Feedback is high quality and consistent across all Question Sets, relating directly to learning points taken from the National Curriculum.

Confidence and resilience building is developed through feedback. Encouraging students to try again takes away the 'fear of failure'.

STEPPED GUIDANCE

We don't just give the correct answer. We give feedback that provides short steps to reach the right answer. This is effective in its simplicity, providing guidance on the action needed to find the correct answer.

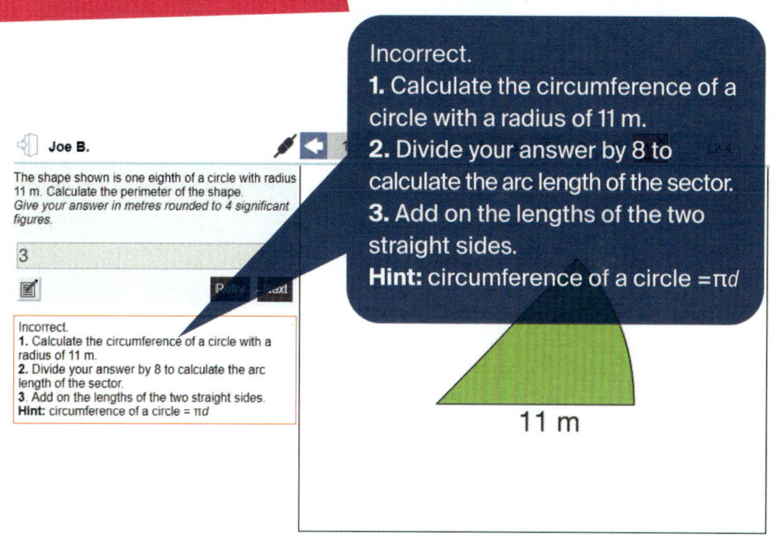

COMMON MISCONCEPTIONS

Common misconceptions are addressed in feedback to encourage pupils to find the correct solution themselves.

REINFORCING DETAIL

Correct-answer feedback provides learners with reinforcing detail that boosts confidence and produces opportunities for deeper learning experiences.

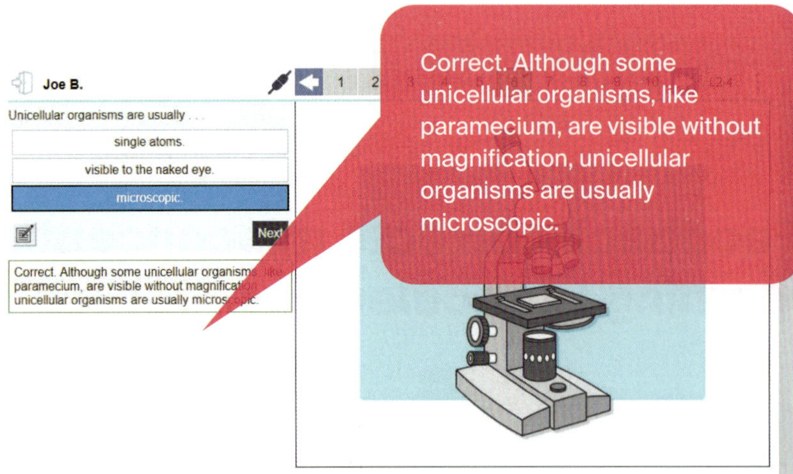

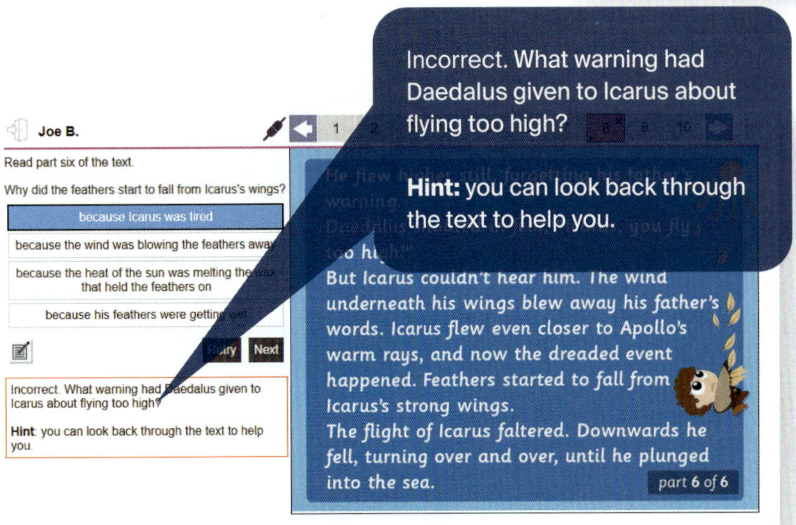

GROWTH MINDSET ENCOURAGEMENT

Feedback that provides information on how to correctly answer the question next time encourages a growth mindset by emboldening pupils to have another go.

SUPPLEMENTARY INFORMATION

Supplementary information in feedback for correct answers develops understanding and nurtures curiosity.

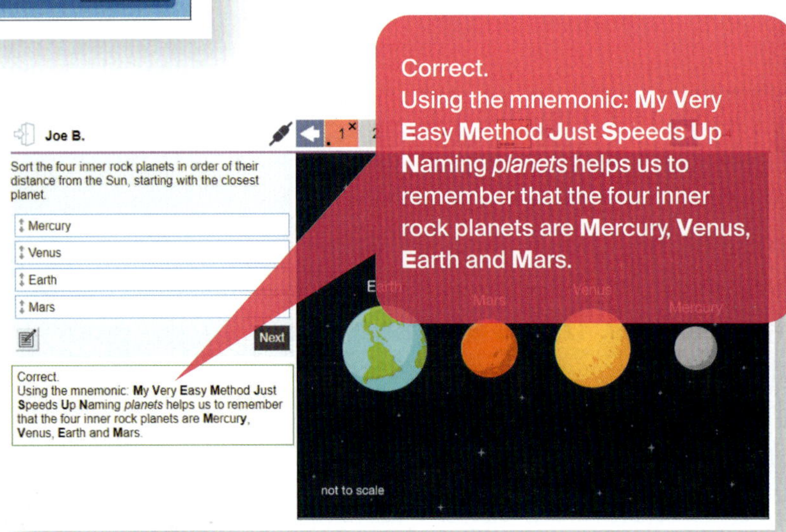

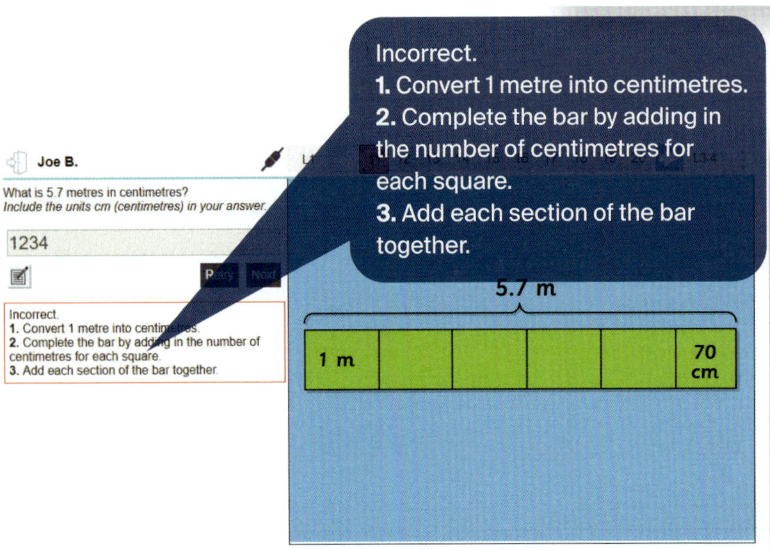

SUPPORTING IMAGERY

Feedback that explains how to use the image provided allows less-secure learners to use pictorial representations rather than abstract understanding.

Mastery Across the Subjects

Learning by Questions (LbQ) is a classroom app filled with curriculum-aligned Question Sets and immediate pupil feedback to supercharge learning. LbQ reduces teacher workloads with automatic marking and real-time lesson reports to verify learning as it happens. LbQ covers KS2 maths, English and science, plus KS3 maths, English, chemistry, physics and biology.

Teachers use LbQ's live data to pinpoint where intervention is needed and why. Pupils use devices such as tablets, to progress through scaffolded Question Sets at their own pace and receive immediate feedback as they answer. Teachers receive live analysis and results are automatically recorded to support assessment and planning and to save time.

Mastery Approach

The mastery approach informs the creation of all of our Question Sets. At LbQ, we believe that mastering a topic or concept is vital. Each pupil's journey to mastery may be different to their peers' journeys, so it is important to us that scaffolding or guiding of learning is built into our Question Sets.

LbQ embraces digital tools to make the mastery approach to learning more achievable than it has ever been before.

Our mastery Question Sets are specifically designed to progress pupils only when they demonstrate a grasp of prerequisites, helping them to master concepts, build a growth mindset and take ownership of their learning. Equally, if a student demonstrates mastery level understanding, progression rules will allow the pupil to move on after a set amount of correct answers, pushing higher ability pupils at pace.

Mathematics

Mastery Question Sets

Maths mastery Question Sets take pupils on a learning journey through four different cognitive levels of a topic: understanding, fluency, reasoning and problem solving. To aid pupil progression, every question in every level provides contextualised feedback and addresses key misconceptions. Maths mastery Question Sets provide students with exposure to problem solving and reasoning questions, but not before previous skills and knowledge have been mastered.

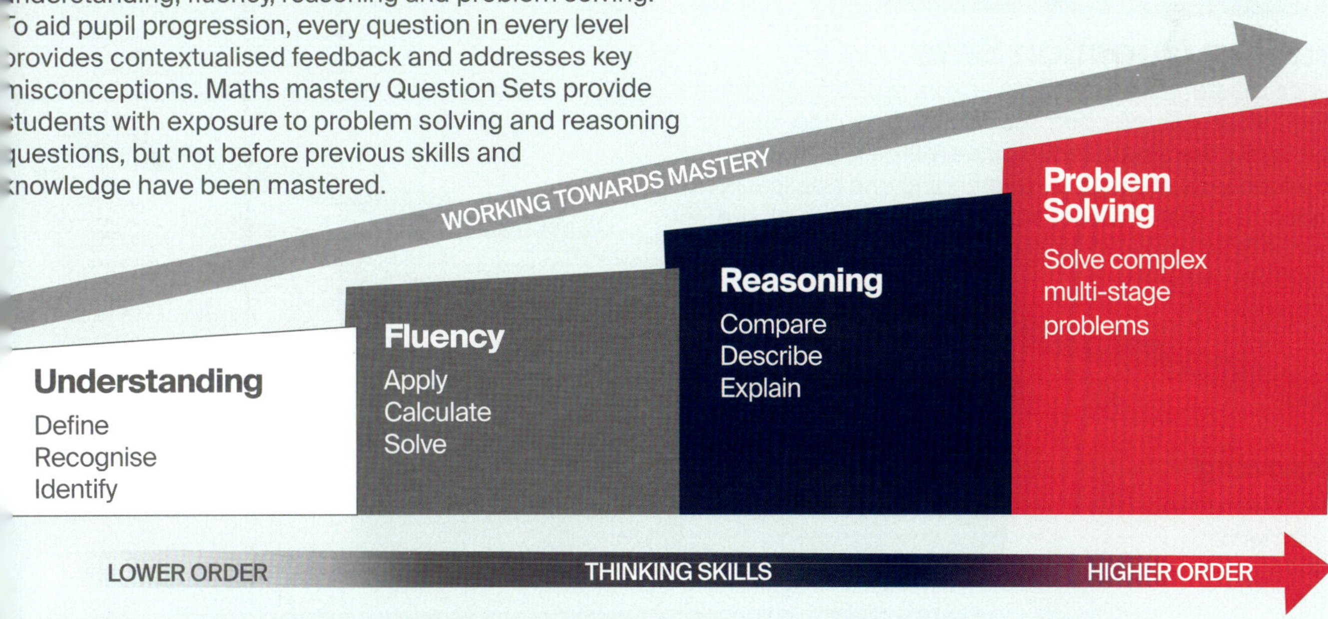

Practice Question Sets

There are elements within the curriculum that require a different approach to learning. Practice Question Sets in maths drill one or two levels of cognition, where pupils recall and apply knowledge rapidly. Such Question Sets are perfect for practising times tables, number bonds and other basic recall of operations needed to progress in maths.

English

Reading Question Sets

This collection of Question Sets for both Key Stage 2 and 3 focuses on developing a wide range of reading skills. At Key Stage 2, skills in retrieval, inference and understanding vocabulary are covered by individual Question Sets on the same text. At Key Stage 3, students are prepared for the skills needed for examinations later in their education through the use of classic literature extracts and original texts.

WORKING TOWARDS MASTERY

Defining
Pre-Read Question Sets
What does...mean?
What do you expect this text to be about?

Recalling
Retrieval Question Sets
How many...?
Which character...?
When did... take place?

Inferring
Inference Question Sets
Why is/does/did...?
Which statement best describes the...?

Analysing
Language Question Sets
What do you think the author meant by using the phrase...?
Why do you think the author uses the word...?
Structure and Features Question Sets
What genre is this text?
Which paragraph explains...?
What is the purpose of...?

LOWER ORDER — THINKING SKILLS — HIGHER ORDER

Grammar, Spelling and Punctuation Question Sets

Key Stage 2 GPS Question Sets are aligned to the National Curriculum. These begin with questions that require pupils to define terminology and understand rules, before moving on to apply rules with contextualised examples.

Key Stage 3 GPS Question Sets build on the skills and knowledge students gained at Key Stage 2. GPS Question Sets at this level are created and labelled by ability to ensure each and every student in your classroom is catered for.

Science

Reading Question Sets

Science mastery Question Sets enable pupils to progress through a carefully stepped sequence of learning. Blocks of questions are structured to guide learners as they move from knowledge, through understanding and application to analysis. Real-world applications are woven through the Question Sets, making learning relevant and relatable.

WORKING TOWARDS MASTERY

Knowing
I can recognise...
I can name...

Understanding
I can use a model to describe...
I understand how...

Applying
I can apply my learning to...
I can suggest ways to...

Analysing/ Evaluating
I can predict...
I can evaluate claims that...

LOWER ORDER — THINKING SKILLS — HIGHER ORDER

Practice Question Sets

LbQ's science practice Question Sets cover areas of the curriculum where repetition of a learning point is essential to ensure that it is thoroughly embedded. They often focus on a skill or area of the curriculum where a weaker grasp of knowledge may become a barrier to future learning, e.g. energy transfer terminology.

Investigation Question Sets

Question Sets support the skills of planning an investigation and analysing results. Pupils will be given opportunities to plan investigations and review data. In KS3, students are provided with further Question Sets that present a mix of exam-style questions and a wider application of the data.

Subscriptions

FULL ACCOUNT

Start a *60 day no-obligation trial of the FULL Account and view affordable pricing at www.lbq.org/trylbq

FULL Account subscription includes:

- connection to pupil devices
- instant pupil feedback
- automatic marking
- real-time lesson analysis
- lesson results saved

+

Self-Paced Questioning

Ad Hoc Questioning

Teach

Use LbQ books to plan and quickly locate Question Sets at www.lbq.org/books

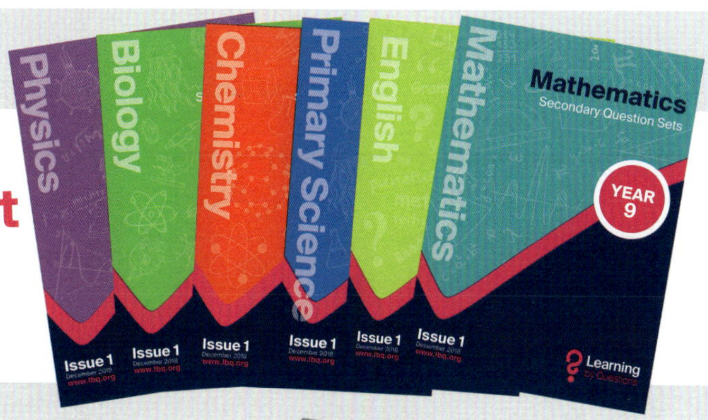

FREE ACCOUNT

Register at www.lbq.org/trylbq

FREE Account includes...

tens of 1,000s of questions with learning feedback for use by teachers on a classroom display.

+

Teach

*Subject to change at discretion of LbQ

Understanding a Question Set

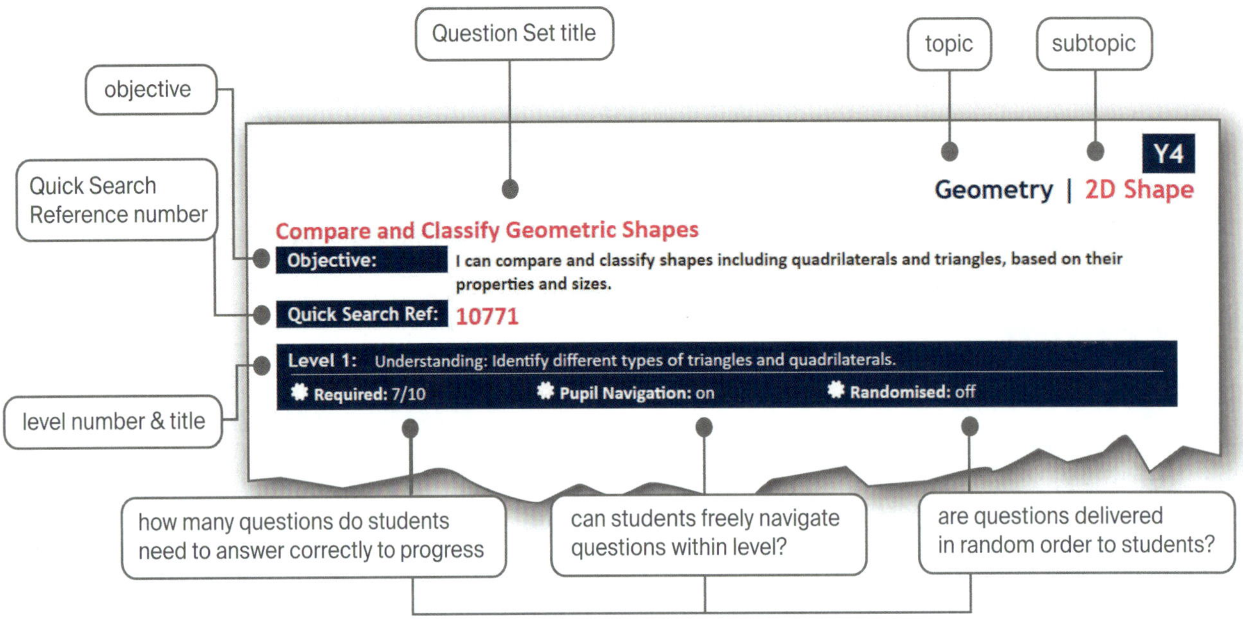

These settings can be adjusted by **adapting** the Question Set.

Understanding a question

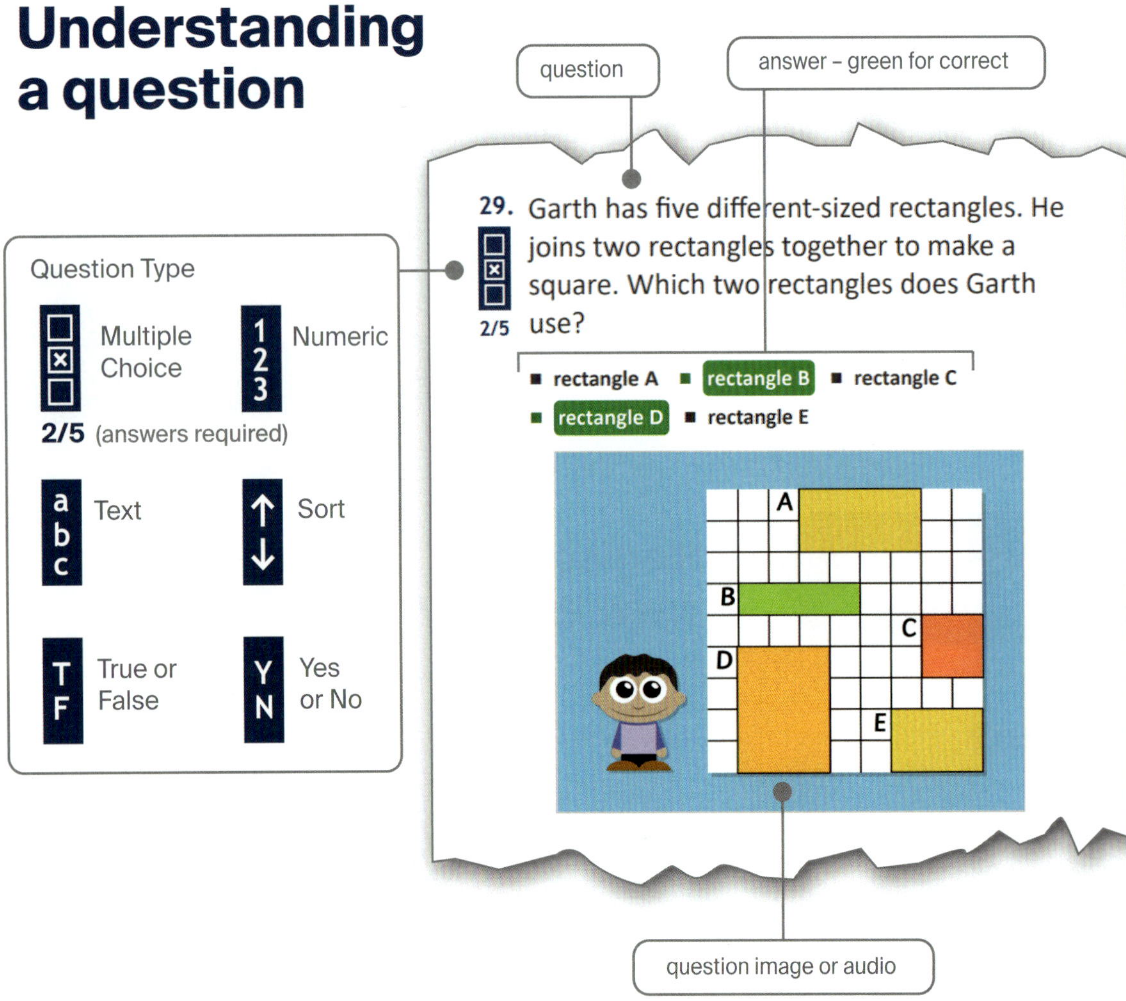

Finding a Question Set From This Book on the LbQ Platform

The **year, topic** and **subtopic** classifications used in this book relate directly to those used online.
The fastest way to find a specific Question Set on lbq.org is via the **Quick Search Reference Number** (e.g. 10771).

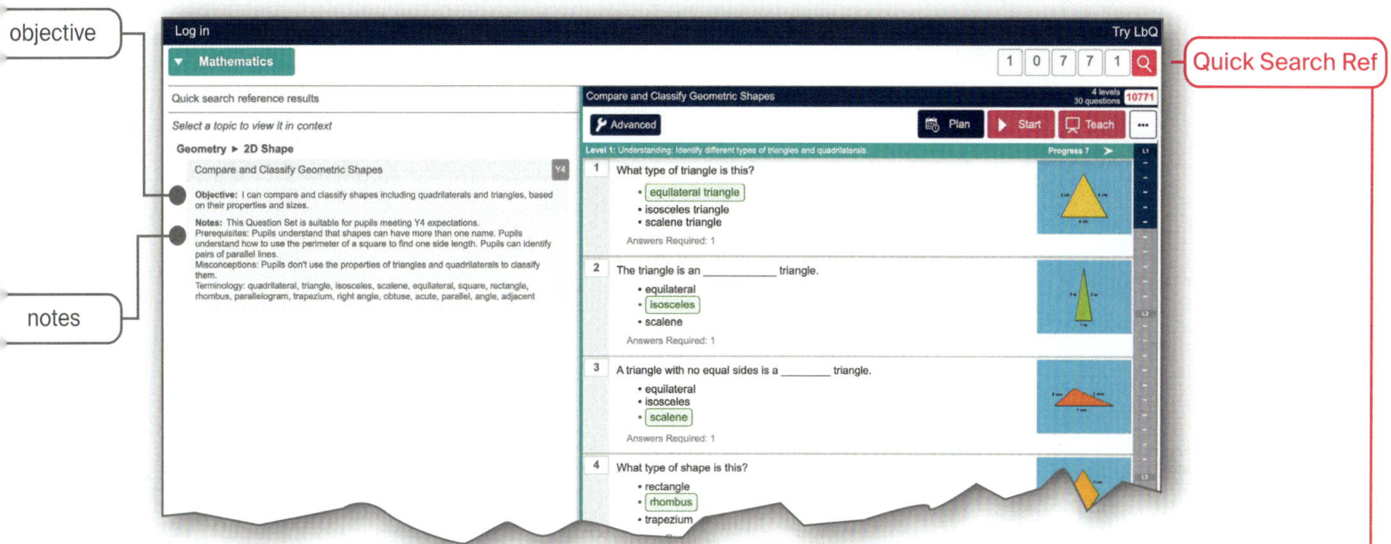

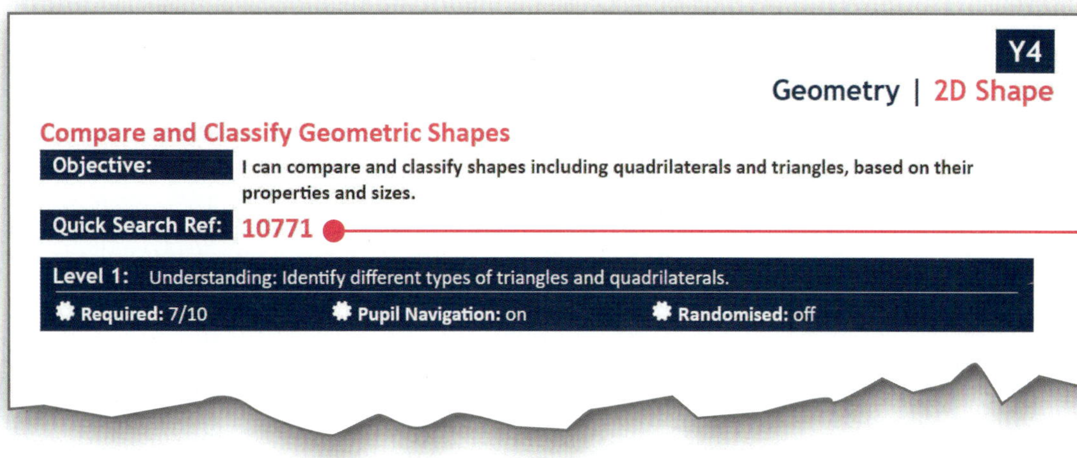
Y4

Geometry | 2D Shape

Compare and Classify Geometric Shapes

Objective: I can compare and classify shapes including quadrilaterals and triangles, based on their properties and sizes.

Quick Search Ref: 10771

Level 1: Understanding: Identify different types of triangles and quadrilaterals.

Required: 7/10 **Pupil Navigation:** on **Randomised:** off

Note: The Question Sets detailed in this book are correct at time of compilation (October 2019) and correspond directly to the Question Sets as published on **www.lbq.org**.

Owing to the nature of **www.lbq.org**, we will from time to time extend, update or modify the Question Sets published, which will at times give rise to discrepancies between this book and the online resources.

Changes and additions to Question Sets will be notified on our website on a regular basis.

Topic Directory Y3/4

Primary Science Y3/4

Plants

Growth
Life Cycles
Nutrition

Investigating What Plants Need to Grow: Plan and Do

Objective: Planning a practical investigation to explore the requirements of plants for life and growth. Investigation plans and materials will need to be provided for the practical work.

Quick Search Ref: 10468

Level 1: Planning an investigation to find out if plants need soil to grow: I can ask relevant scientific questions; I can make a prediction with a reason.

Required: 11/11 **Pupil Navigation:** on **Randomised:** off

1. These children have been discussing what it would be like to live on the planet Mars. What could they investigate to find out if people will be able to grow food on Mars?
1/3

- What types of plants can grow without much water?
- Can plants grow without soil?
- Can plants grow in cold temperatures?

2. Working safely: Select the most important safety rule for working with plants.
1/3

- Wash your hands after touching plants.
- Wear safety glasses.
- Put thermometers in the centre of the table.

3. You are going to plant some seeds to investigate if plants need soil to grow. Which variable will you change in your investigation?
1/4

- amount of water used
- material you plant the seeds in
- size of plant pots
- where you put the plant pots

4. Which variable could you measure to find out if the plants are growing?
1/4

- heights of plants
- number of new leaves
- time taken to grow a new leaf
- any of these

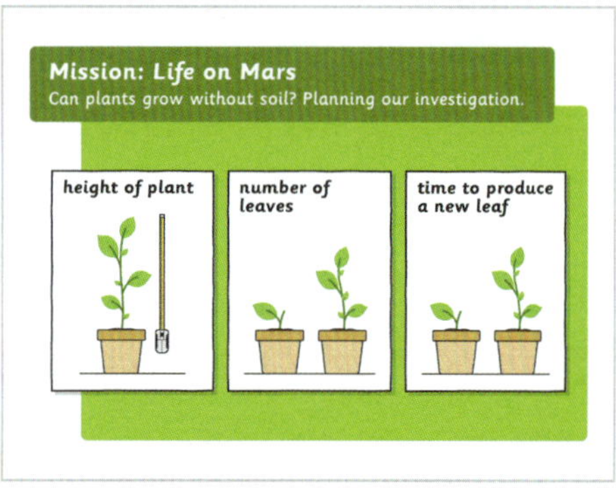

5. Which scientific question are you investigating?
1/3

- What happens to the number of leaves on the plants when you change the material you plant the seeds in?
- What happens to the heights of the plants when you change the size of the plant pots?
- What happens to the heights of the plants when you change the material you plant the seeds in?

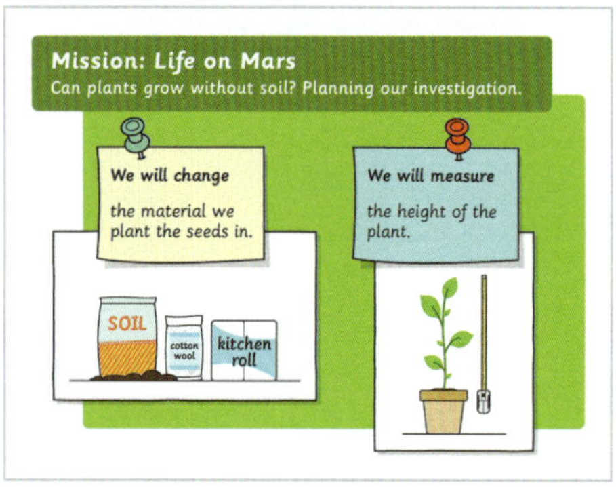

Level 1 *continued*

6. Which **three variables** should you keep the same to make it a fair test?

3/4

- **amount of water used**
- **size of plant pots**
- **where you put the plant pots**
- **material you plant the seeds in**

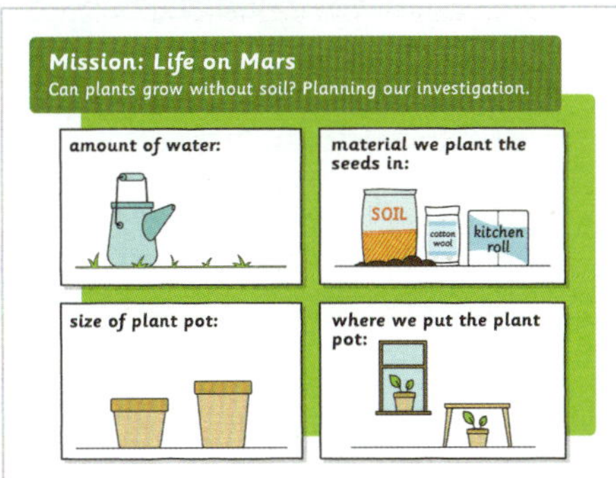

7. What do you think will happen to the heights of the plants growing in the different materials?

1/4

Complete the prediction:
The tallest plant will be the one growing in . . .

- soil.
- cotton wool.
- kitchen paper.
- nothing.

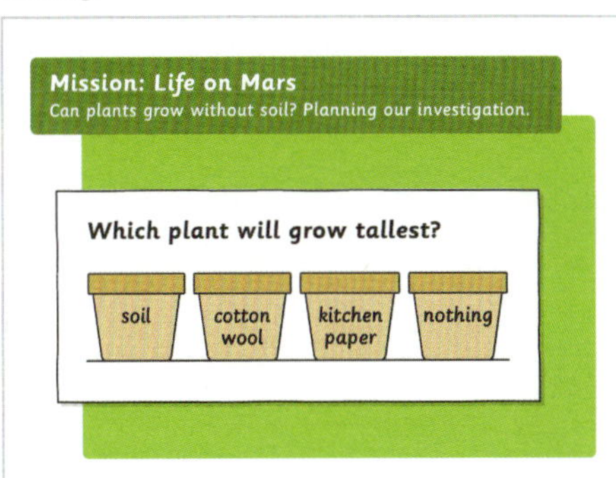

8. Which piece of equipment will you use to measure the heights of the plants?

1/4

- **stopwatch**
- **measuring cylinder**
- **ruler**
- **thermometer**

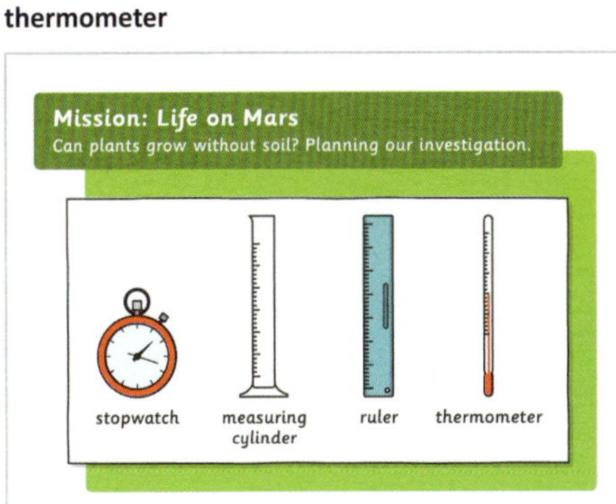

9. How many seeds will you plant in each pot?

1 2 3

- Open question, no set answer

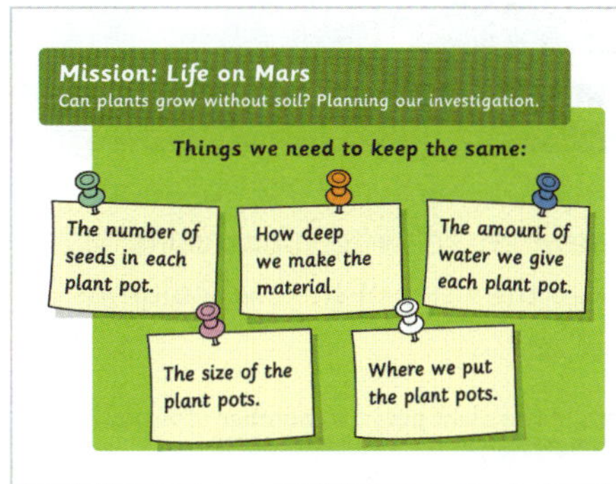

10. How deep will you fill the pots with each material?

a b c

- Open question, no set answer

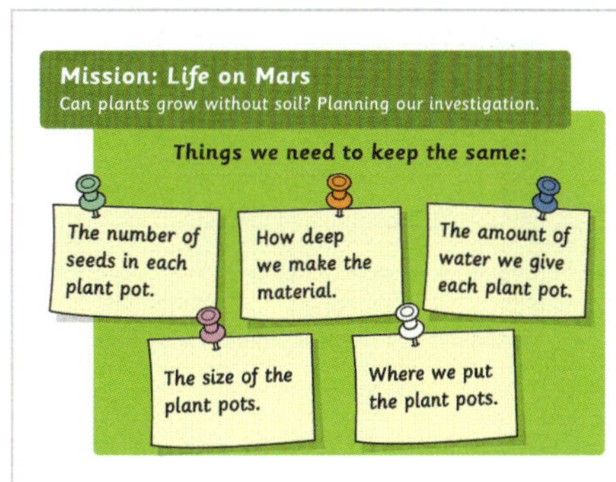

11. What else do you need to decide?

a b c

- Open question, no set answer

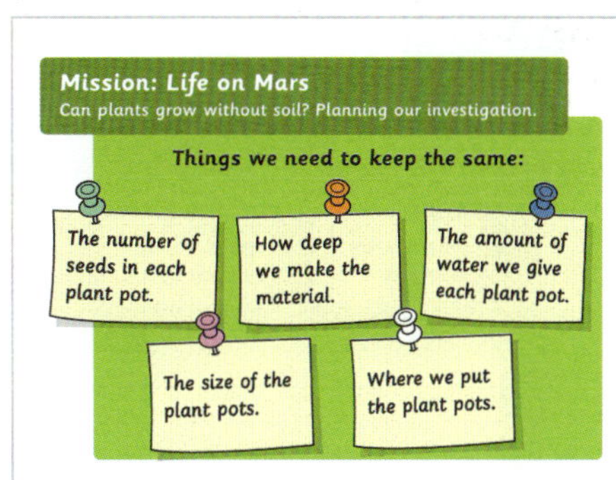

Level 2: Measuring and recording: I can make careful and accurate observations; I can record data in a table.

Required: 5/5 **Pupil Navigation:** on
Randomised: off

12. Before you start your investigation, you need somewhere to write your results. What should the titles of the two columns of your table be?

2/6

- size of the plant pots
- number of leaves
- **material you planted the seeds in**
- amount of water used
- time taken to grow a new leaf
- **height of the plant**

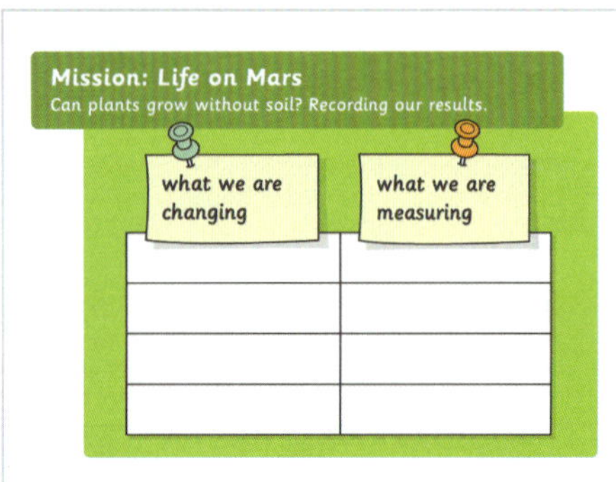

13. What units will you use to measure the heights of the plants?
Remember to put these units in the column title.

1/3

- **millimetres (mm)**
- centimetres (cm)
- metres (m)

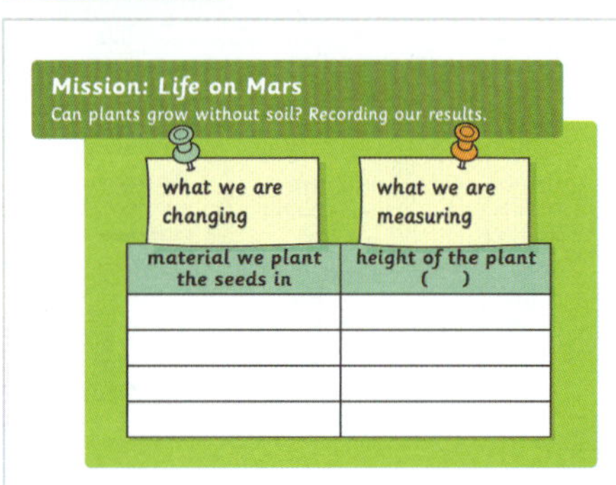

14. What else do you need to put in your table?

1/4

- where you put the plant pots
- the amount of water used
- **the name of each material**
- the size of the plant pots

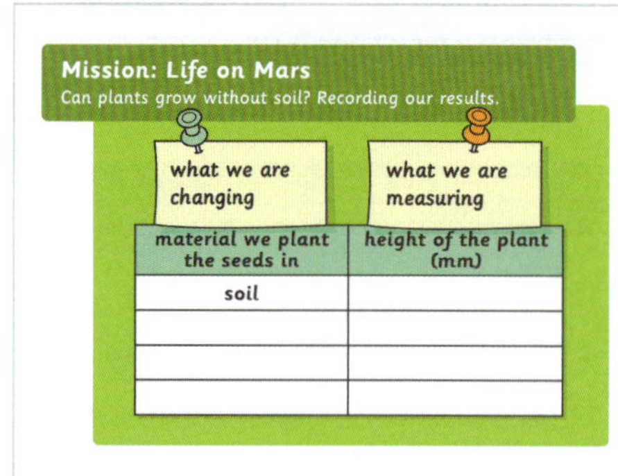

15. If you have more than one plant in each pot, which plant should you measure and record the height of?

1/4

- all of them
- **the tallest one**
- the shortest one
- the greenest one

16. How long will you leave the plants to grow before you measure them?

- Open question, no set answer

Investigating What Plants Need to Grow: Review

Objective: Presenting and reviewing results from a practical investigation to explore the requirements of plants for life and growth. Results tables from the investigation and graph paper will need to be available.

Quick Search Ref: 10469

Level 1: Reviewing the plan for the investigation to find out if plants need soil to grow: I can ask relevant scientific questions; I can make a prediction with a reason.

Required: 7/7 **Pupil Navigation:** on **Randomised:** off

1. You carried out an investigation to find out if we could grow food on Mars. What did you investigate?

1/3

- What types of plants can grow without much water?
- **Can plants grow without soil?**
- Can plants grow in cold temperatures?

2. You planted some seeds to investigate if plants need soil to grow.
Which variable did you change in your investigation?

1/4

- amount of water used
- **material you plant the seeds in**
- size of plant pots
- where you put the plant pots

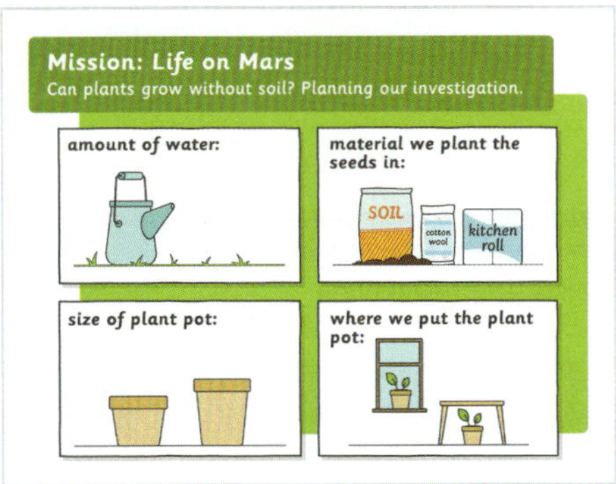

3. Which variable did you measure to find out how well the plants were growing?

1/3

- number of new leaves
- time taken to grow a new leaf
- **heights of the plants**

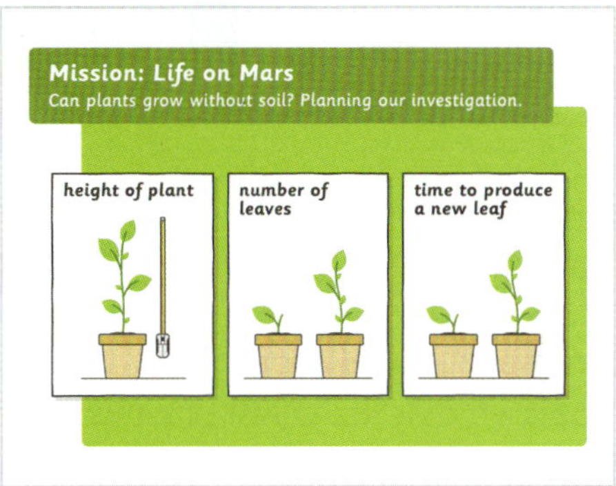

4. Which scientific question did you investigate?

1/3

- What happens to the number of leaves on the plants when you change the material you plant the seeds in?
- What happens to the heights of the plants when you change the size of the plant pots?
- **What happens to the heights of the plants when you change the material you plant the seeds in?**

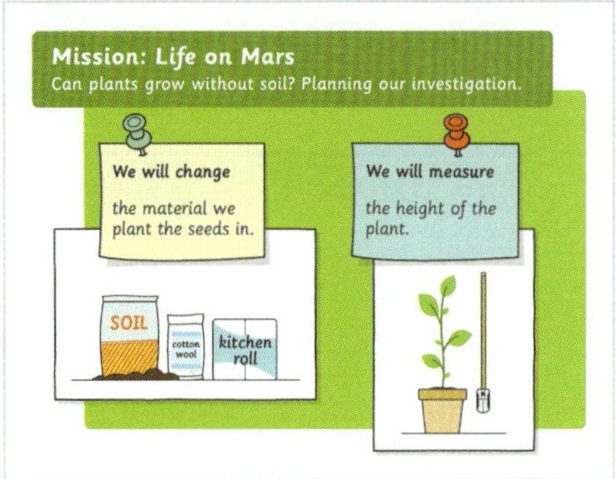

Level 1 *continued*

5. Which **three variables** did you keep the same to make it a fair test?

3/4

- amount of water used
- size of plant pots
- where you put the plant pots
- material you plant the seeds in

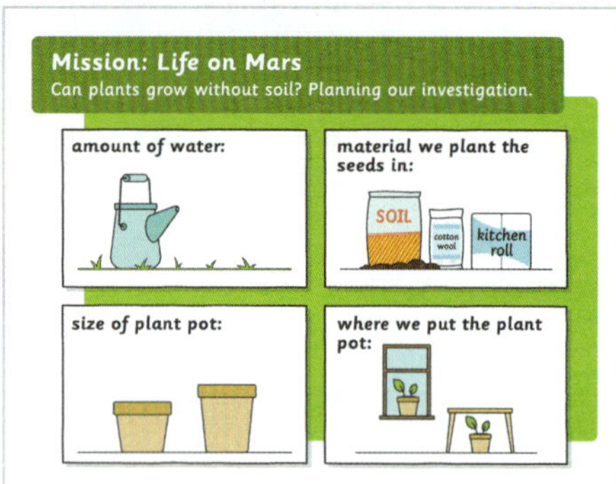

6. What did you think would happen to the heights of the plants growing in the different materials?
Complete the prediction:
I predicted that the tallest plant would be the one growing in . . .

1/4

- soil.
- cotton wool.
- kitchen paper.
- nothing.

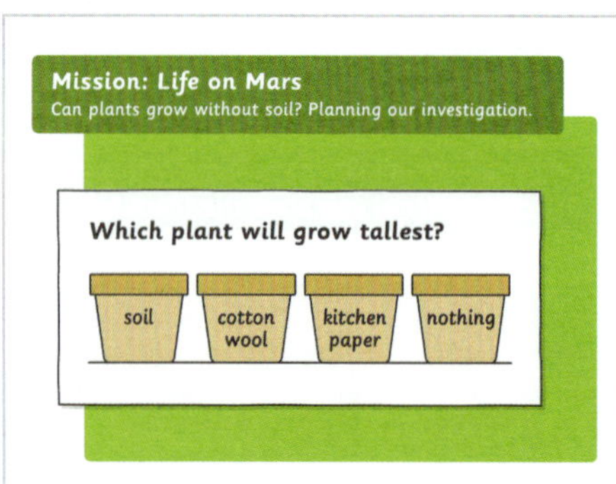

7. Before starting your investigation, you drew a results table. What are the titles of the two columns of your table?

2/6

- size of the plant pots
- number of leaves
- material you planted the seeds in
- amount of water used
- time taken to grow a new leaf
- height of the plant

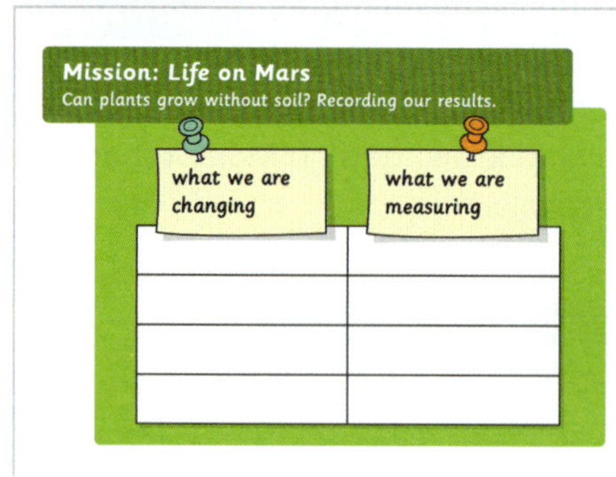

Level 2: Presenting: I can present data in a bar chart.

Required: 5/5 **Pupil Navigation:** on
Randomised: off

8. What would be the best way to show the results of your investigation?

1/3

- in a tally chart
- in a bar chart
- in a pictogram

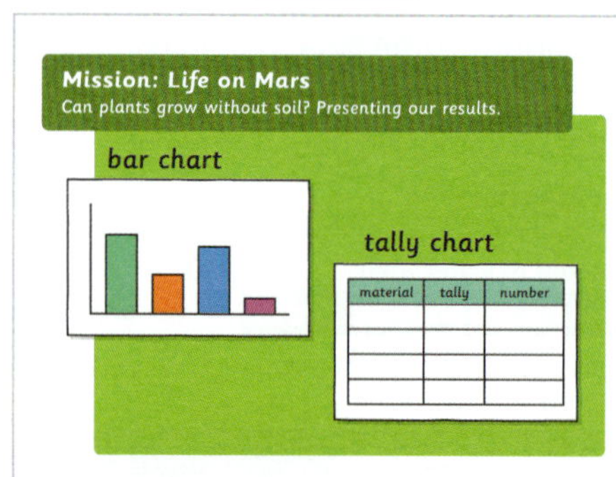

9. Which variable goes on the horizontal axis?

1/2

- material you planted the seeds in
- height of the plant (mm)

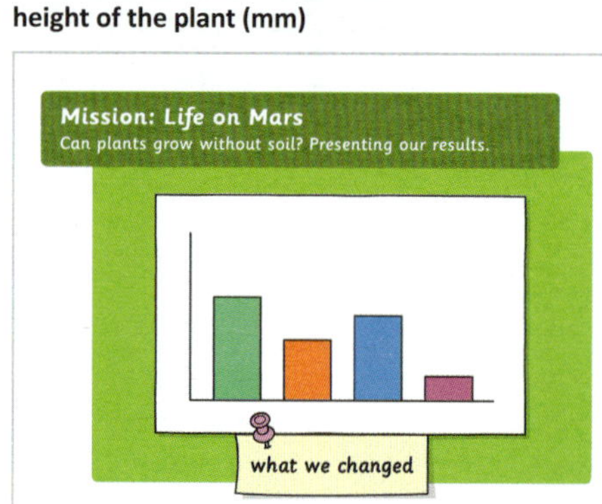

Level 2 *continued*

10. Which variable goes on the vertical axis?

1/2

- material you planted the seeds in
- **height of the plant (mm)**

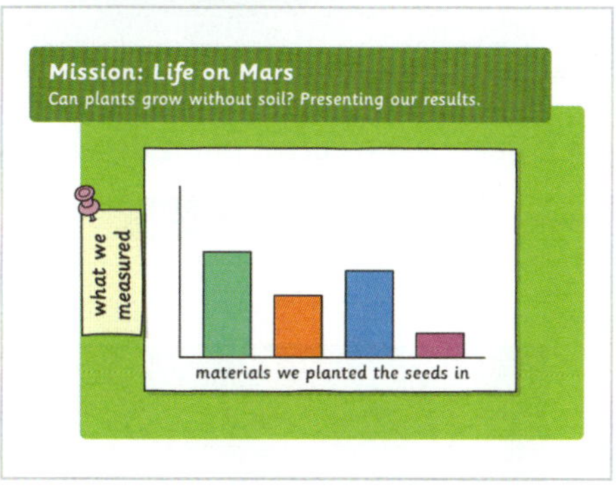

11. How many bars will you draw on your bar chart?

1 2 3

- Open question, no set answer

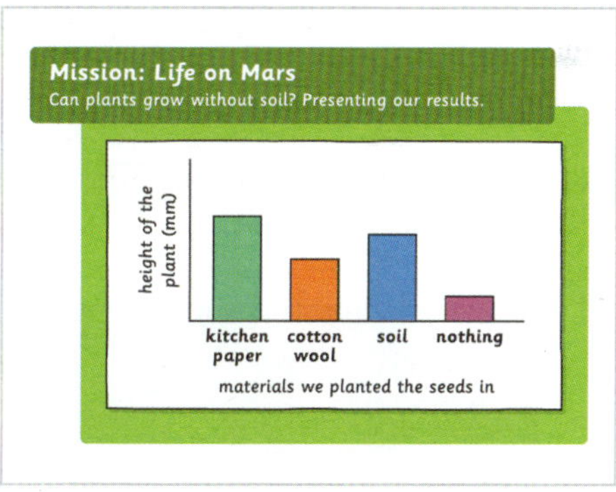

12. Complete the title for your graph:
Graph to show . . .

1/3

- the heights of the plants in the investigation
- the materials used in the investigation
- **the heights of plants depending on the material they were growing in**

Level 3: Analysing and reviewing: I can identify differences, similarities and changes related to an enquiry; I can draw conclusions and suggest improvements.

Required: 5/5 **Pupil Navigation:** on
Randomised: off

13. What does the tallest bar on your bar chart show you?

1/4

- which material the shortest plant grew in
- which material the healthiest plant grew in
- which material the plant with the most leaves grew in
- **which material the tallest plant grew in**

14. Did your results agree with your prediction? Explain how you know.

a b c

- Open question, no set answer

15. Was there anything you found difficult to control in your investigation? Explain how this might have stopped it being a fair test.

a b c

- Open question, no set answer

16. How would you check your results?

1/3

- repeat the investigation for one of the materials
- **repeat the whole investigation**
- repeat the investigation using different seeds

17. What did your investigation tell you about whether or not we could grow food on Mars?

a b c

- Open question, no set answer

Plant Parts and their Functions

Objective: I can identify and describe the functions of different parts of flowering plants: roots, stem/trunk, leaves and flowers.

Quick Search Ref: 10461

Level 1: Knowledge: I can identify different parts of flowering plants and trees.

Required: 6/6 **Pupil Navigation:** on **Randomised:** off

1. Which part of the plant does label A point to?

1/5

- stem
- **roots**
- flower
- leaf
- fruit

2. Sort the plant parts into the order shown in the diagram. The part that matches label 1 goes at the top of the list.

- **flower**
- **stem**
- **leaves**
- **roots**

3. The needles are which part of the pine tree?

1/4

- roots
- stem
- **leaves**
- flowers

4. Which part of the carrot plant do we eat?

1/4

- flowers
- stems
- leaves
- **roots**

5. Which of these vegetables is a **stem**?

1/5

- broccoli
- carrot
- tomato
- **celery**
- peas

6. The red part of a rhubarb plant is its _____.

- **stem.**
- **stem**

Level 2: Understanding: I can describe the function of different parts of flowering plants and trees.

Required: 5/5 **Pupil Navigation:** on
Randomised: off

7. A green plant makes its nutrition in its ______.

- leaf
- leaves
- stems
- leef
- root
- flower
- leafs
- leefs
- roots
- stem
- flowers

A Flowering Plant

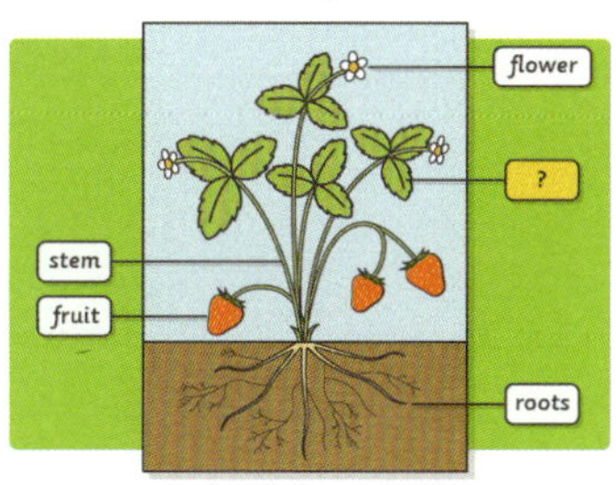

8. The function of the flower is to . . .

1/4

- provide nutrition (food).
- support the plant.
- bring water into the plant.
- reproduce (make more plants).

Flowers

9. _ _ _ _ _ grows from part of the flower and contains the seeds for new plants.

- fruit
- frute
- fruits
- froot
- friut

Stages in Fruit Development

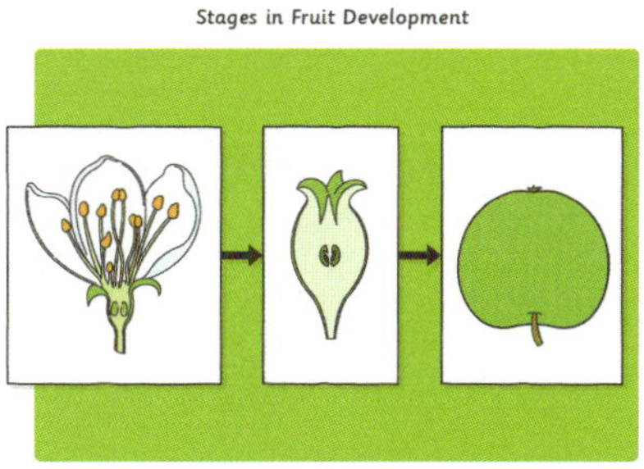

10. What are the **two main things** roots do?

2/4

- take water into the plant
- support the leaves
- make the plant's food
- hold the plant in the ground

A Flowering Plant

11. Water reaches the flowers and leaves through the _ _ _ _.

- stems
- stem.
- stem
- stems.

A Flowering Plant

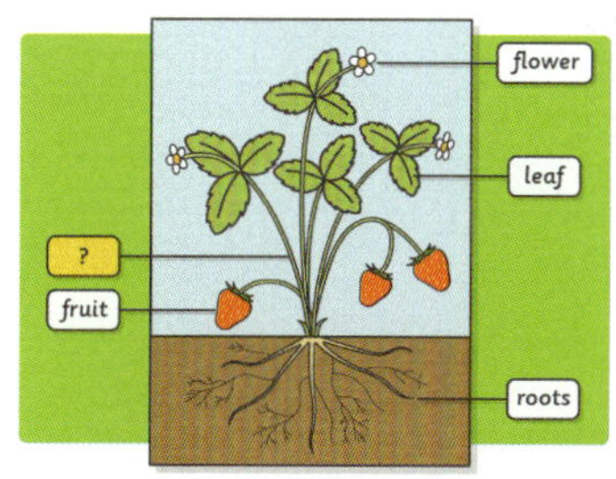

Level 3: Application: I can apply my knowledge and understanding of plant parts to the world around me.

Required: 2/2 **Pupil Navigation:** on
Randomised: off

12. Why would a plant need deep tap roots?

2/4

- to anchor them firmly so they don't fall over in strong winds
- to take in nutrients from shallow soil
- to reach water far below the surface in dry environments
- to be able to quickly take in surface water from rainfall

A Flowering Plant

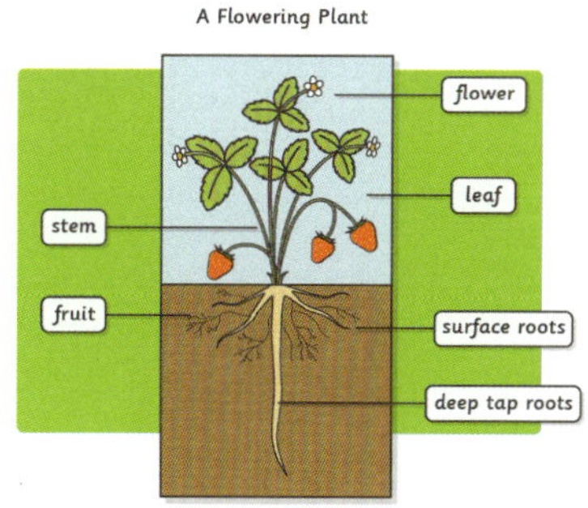

Level 3 *continued*

13. If plants create food in their leaves, why do farmers put fertiliser on their fields?

1/3

- **The fertiliser helps the plants get food at night when there isn't any sunlight.**
- **The farmers use fertiliser to kill weeds.**
- **The fertiliser adds additional nutrients to the water that is taken in by the plants' roots.**

Farmer Adding Fertiliser to the Soil

Plant Parts Vocabulary

Objective: I can spell and define scientific words for plant parts.

Quick Search Ref: 10445

Level 1: Knowledge: I can spell and define scientific words for plant parts.

Required: 7/7 **Pupil Navigation:** on **Randomised:** off

1. What word completes the plant fact?

abc

The _______ holds the plant in the ground and takes up water from the soil.

- stem
- routes
- **roots**
- **root**
- leafs
- flower
- route
- leaf
- leaves

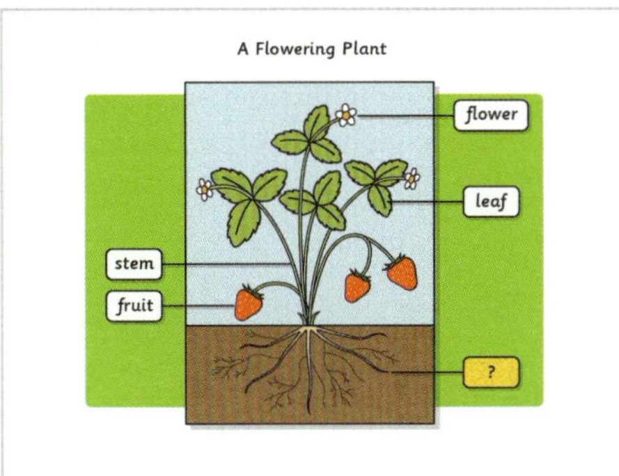

2. Select **three** functions of the stem.

3/5

- Makes seeds.
- **Holds the plant upright.**
- Anchors the plant.
- **Supports the leaves.**
- **Moves water around the plant.**

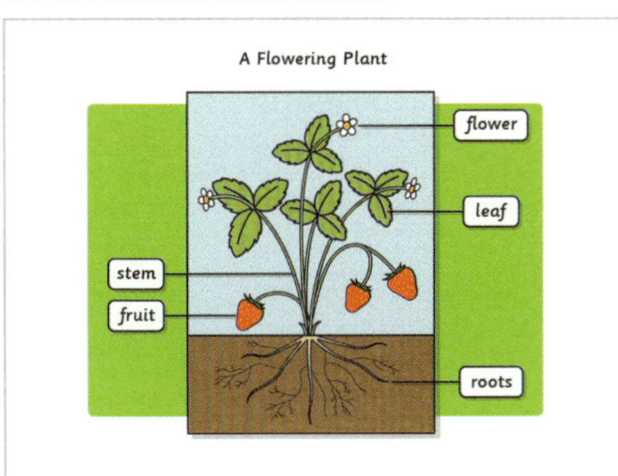

3. Which part of the plant is where seeds are made to form new plants?

1/4

- the stem
- the roots
- the leaves
- **the flower**

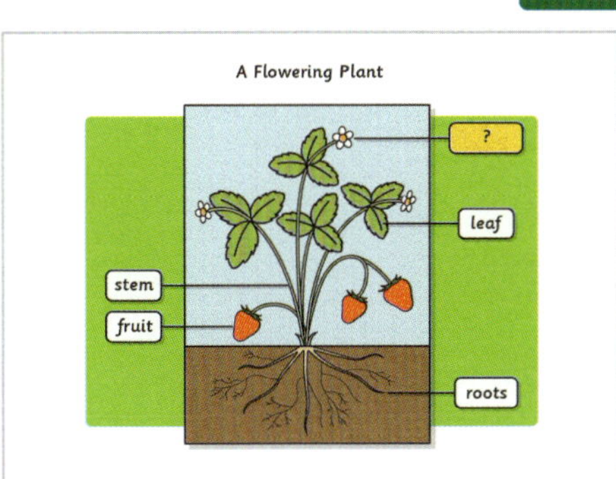

4. Green plants make their own food in their _ _ _ _ _ _.

abc

- **leaves.**
- **leaf.**
- **leaves**
- **leaf**
- leafs
- stem
- roots
- laeves
- flower

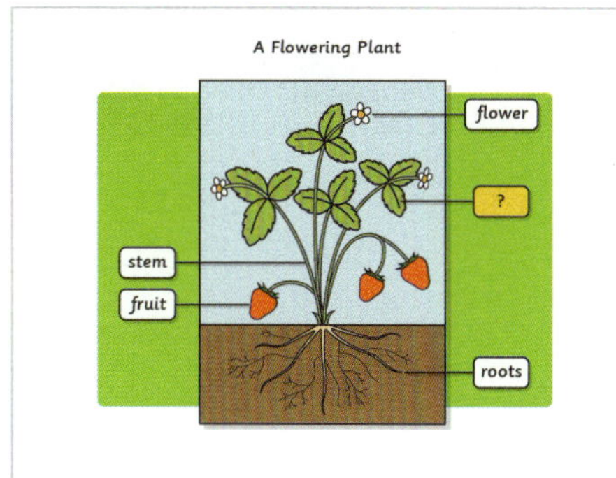

5. Germination is when . . .

1/3

- **a seed starts to grow.**
- a new seed is made.
- a seed is spread from the parent plant.

6. What is the name of the plant part that contains the seeds?

abc

- **fruit**
- **the fruit**
- **fruit.**
- froot
- **the fruit.**
- frute
- the frute
- the froot

Level 1 *continued*

7. Veins in the leaves . . .

☐ ☒ ☐ 1/3

- contain seeds.
- make pollen.
- transport water and nutrients.

What Plants Need to Grow

Objective: I can describe the requirements of plants for life and growth and how they vary from plant to plant.

Quick Search Ref: 11258

Level 1: Knowledge: I can describe the needs of plants for growth.

Required: 9/10 **Pupil Navigation:** on **Randomised:** off

1. Plants need water, air, light and nutrients to grow. Plants take in water through their ______.

1/3

- leaves
- **roots**
- stems

2. Animals get the energy they need from ______.

- **food.**
- **food**

3. Plants make their own ______ in their leaves using light, air and water.

- **food**
- **sugar**
- **glucose**
- energy

4. Plants need air, light and water to make their own food. Plants make their food in their ________ .

- **leaves.**
- **leaves**
- roots
- leaf
- leafs

5. Plants make their own food in their leaves using light, air and ______.

- **water.**
- **water**
- nutrients

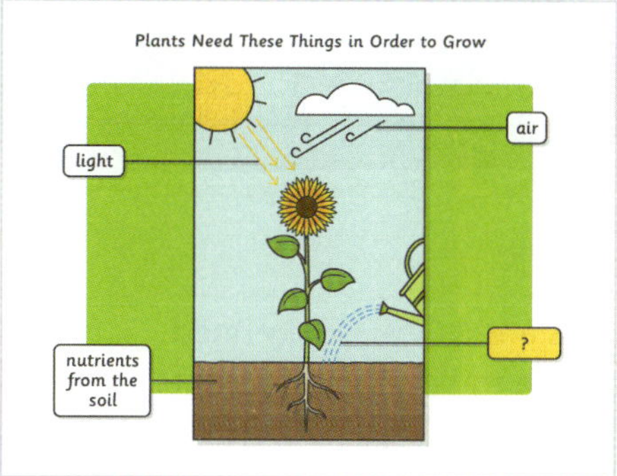

6. Plants make their own food in their leaves using ______, air and water.

- **sunlight**
- **light**
- sun
- the sun

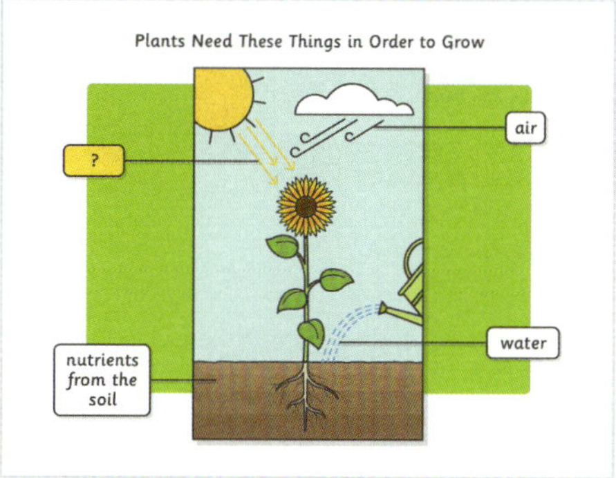

Level 1 *continued*

7. The water that plants take in through their roots contains small amounts of extra nutrients that . . .

1/2

- give them energy.
- help them to become strong and healthy.

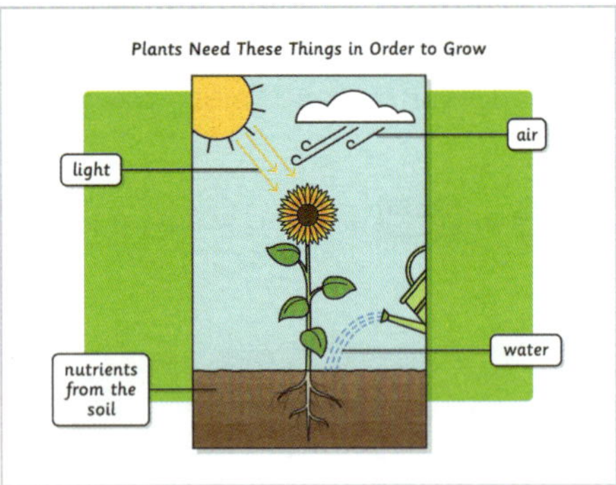

8. Seeds usually start to grow into plants (germinate) when the weather starts to get ______ during spring or summer.

1/3

- brighter
- warmer
- colder

photo by USDA

9. Seeds and bulbs usually need _____ and warmth to develop into plants.

1/2

- water
- sunlight

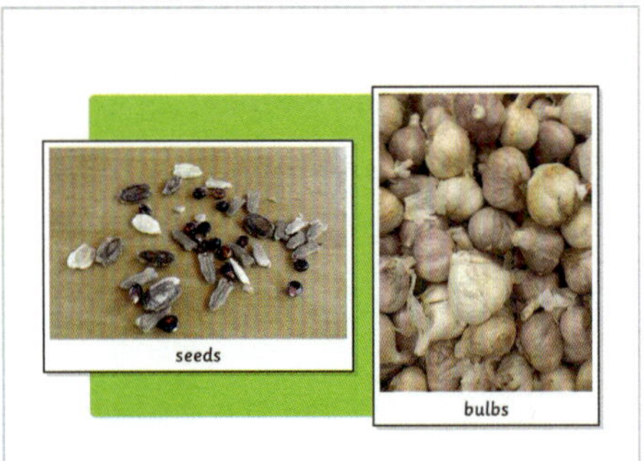

seeds by Karen and Brad Emerson

10. Plants make their own food in their ______ using light, air and water.

- roots
- leaves
- leaf
- leafs

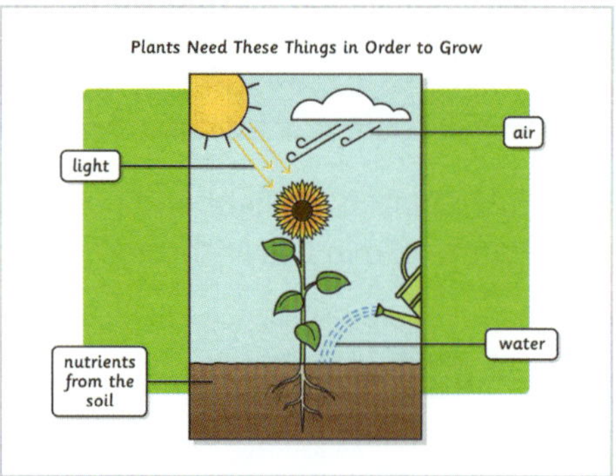

Level 2: Understanding: I can describe the needs of different plants.

Required: 5/5 **Pupil Navigation:** on
Randomised: off

11. Deserts get very little rain. Tia, Tom and Theo are talking about whether or not plants that live in deserts need water. Who is correct?

1/3

- Tia
- Tom
- Theo

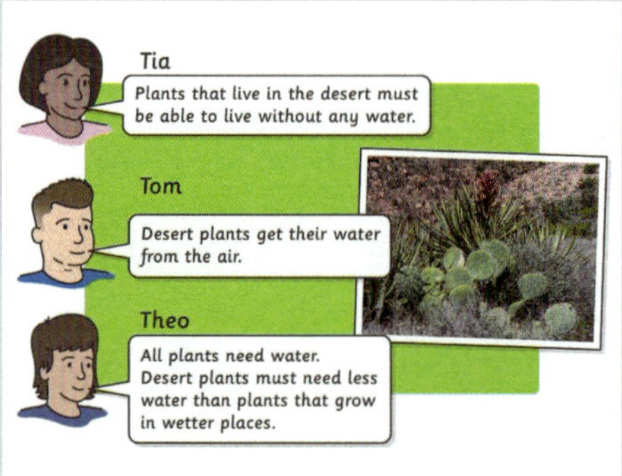

12. Venus flytrap plants grow in boggy areas where the soil doesn't have many nutrients. Dylan, Mo and Eva are discussing why Venus flytraps catch flies. Who is correct?

1/3

- Dylan
- Eva
- Mo

Level 2 *continued*

13. Katie plants seeds in two pots of soil. She puts one pot in the light on the windowsill and the other in a dark cupboard. Katie waters the seeds with the same amount of water each day. After a week, she sees seedlings in both pots. Katie has found that seeds ________________ to germinate and grow into seedlings.

1/4

- need water
- don't need water
- need light
- don't need light

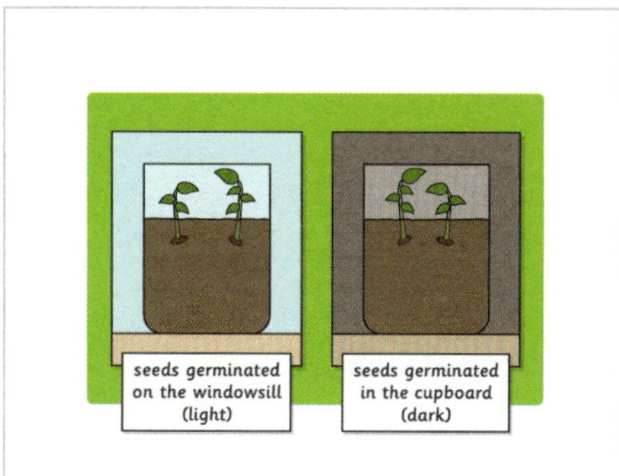

14. Katie has two pots containing identical seedlings. She puts one pot in the light on the windowsill and the other in a dark cupboard. What will the seedlings that are grown in the dark look like after two weeks?

1/3

- They will be paler and weaker than the seedlings grown on the windowsill.
- They will look the same as the seedlings grown on the windowsill.
- They will look stronger and greener than the seedlings grown on the windowsill.

15. If there are too many plants trying to grow in one area, some will not be able to make enough food. Select **two** reasons for this.

2/4

- There might not be enough nutrients in the soil for all the plants.
- There might not be enough water for all the plants.
- People might pick the flowers.
- Bigger plants might stop the sunlight from reaching smaller plants.

Level 3: Application: I can explore the needs of different plants for survival.

Required: 1/3 **Pupil Navigation:** off
Randomised: on

16. Honeysuckle is a plant that often grows in woods. It seems to climb up trees and other tall plants. Use what you know about what plants need to grow well to explain why honeysuckle climbs up other plants.

a b c

- Open question, no set answer

Level 3 *continued*

17. Plants don't usually grow well in very cold places where the water is frozen. Use what you know about what plants need to grow well to explain why plants rarely grow in very cold places.

a b c

- Open question, no set answer

Marmolada Glacier

18. When the children in Mr Grey's class visit a park near their school, they notice that there aren't many plants growing underneath the biggest trees. Explain why you think other plants might not get what they need to grow in the ground around the trees.

a b c

- Open question, no set answer

A Tree in a Park

Life Cycle of a Flowering Plant

Objective: I know the part that flowers play in the life cycle of flowering plants, including pollination, seed formation and seed dispersal.

Quick Search Ref: 10500

Level 1: Knowledge: I can describe the life cycle of a flowering plant.

Required: 6/9 **Pupil Navigation:** on **Randomised:** off

1. Starting with 'The seed germinates.', arrange the stages of the life cycle of a flowering plant from 1 to 6.

- **The seed germinates.**
- **The seedling grows into a young plant.**
- **The plant flowers.**
- **The flower is pollinated.**
- **The fruit develops.**
- **The seeds are dispersed.**

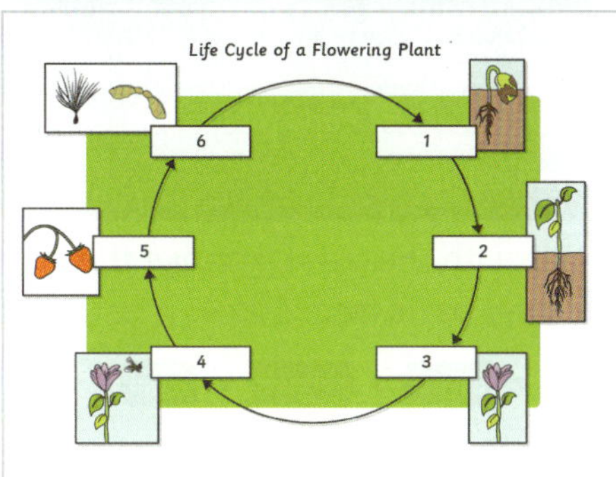

2. What is it called when pollen is transferred from the male part of a flower to the female part of a flower?

1/3

- germination
- fertilisation
- **pollination**

3. Seed dispersal is when . . .

1/4

- pollen joins with an ovule.
- **seeds are spread away from the parent plant.**
- seeds start to grow into new plants.
- pollen is transferred from the male part of a flower to the female part of another flower.

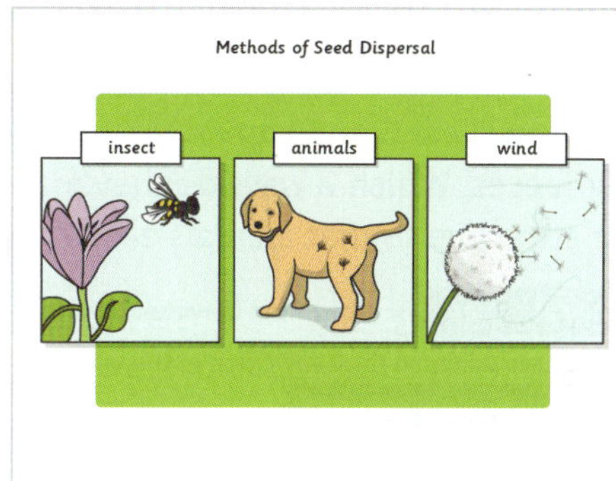

4. Flowers can be pollinated by insects, some other animals and the _ _ _ _ .

- **wind.**
- **wind**

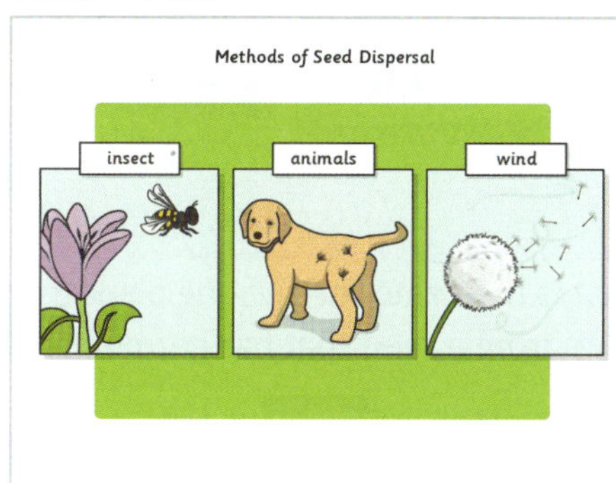

5. Where are the seeds formed in the flower?

1/5

- anther
- style
- **ovary**
- stigma
- filament

Level 1 *continued*

6. Which picture shows germination?

1/4

- picture A
- picture B
- picture C
- **picture D**

7. Flowering plants often grow fruit so that animals will eat it and spread the seeds with their faeces. Which **2** ways do they make the fruit attractive so that animals eat it?

2/4

- **They are sweet.**
- They float in water.
- They have wings so they will catch in the wind.
- **They are brightly coloured.**

8. What is the name for a young plant that grows from a seed that has germinated?

1/4

- bud
- cutting
- **seedling**
- bulb

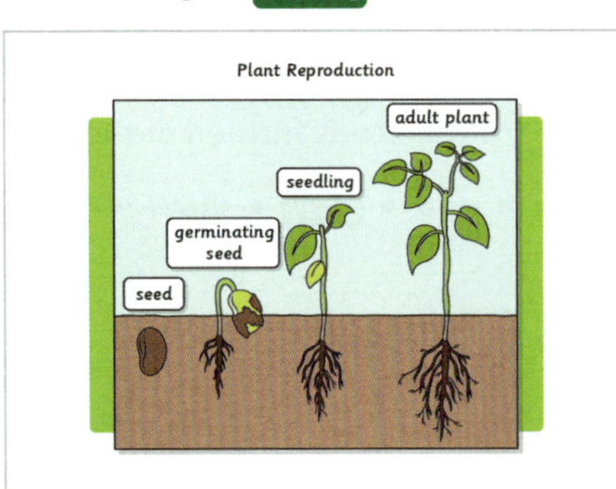

9. The fruit is formed from the female part of the flower. Is this true or false?

1/2

- **True**
- False

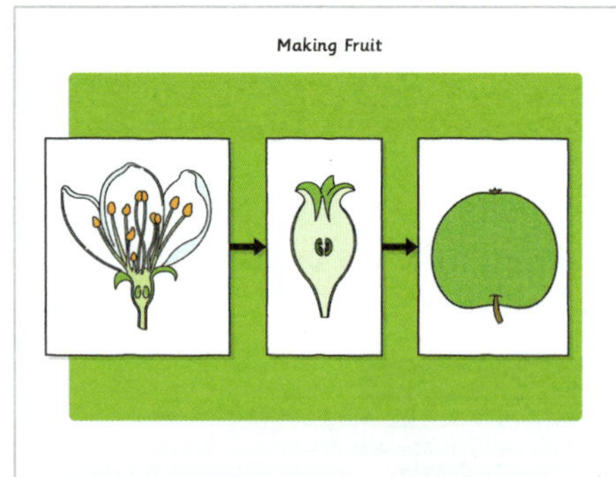

Level 2: Understanding: I can describe the stages of the plant life cycle and the importance of flowers.

Required: 5/5 **Pupil Navigation:** on
Randomised: off

10. Starting with the petal, arrange the labels for the flower parts from 1 to 4.

- **petal**
- **carpel**
- **stamen**
- **sepal**

11. Where is the pollen made in the flower?

1/4

- ovary
- **anther**
- petal
- filament

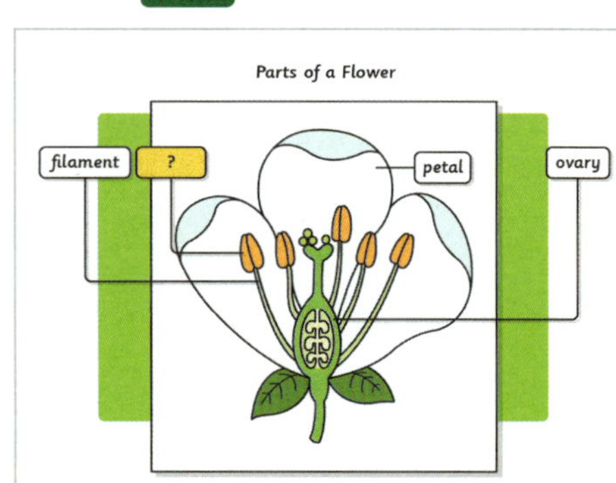

Level 2 *continued*

12. What are **three features** of some wind pollinated flowers.

3/5

- bright petals
- long stamens hanging outside the flower
- large amounts of pollen produced in the anthers
- strong scent
- feathery stigmas hanging outside the flower

13. Select **three features** of some seeds that are dispersed by the wind.

3/5

- dull coloured flowers
- a part that acts like a parachute
- a part that acts like a wing
- lots of pollen
- small, light seeds

14. Jenna has picked all the flowers from one of her plants. What will happen to the plant without any flowers?

1/3

- The plant will die.
- The plant won't make any seeds for new plants.
- The seeds for new plants won't be spread to new places.

Level 3: Application: I can apply my knowledge and understanding of plant reproduction to the world around me.

Required: 3/3 **Pupil Navigation:** on
Randomised: off

15. What problems might there be if seeds grow too close to the parent plant?

2/4

- The seeds might not germinate.
- The plants might not be able to pollinate.
- The new plants might not get light because they are in the shade of the taller parent.
- There might not be enough water for all the plants.

16. Jo wants to protect her strawberry plants from insects, so she covers some of them in plastic before the flowers grow. Will there be any strawberry fruits on the covered plant? *Select yes or no and a reason.*

1/5

- yes
- no
- There will be more strawberries because the insects won't eat them.
- The strawberries will be bigger on the covered plants because they will be protected from the weather.
- There won't be any strawberries because the flowers won't get pollinated.

Level 3 *continued*

17. In the Millenium Seed Bank, scientists have stored seeds from all of Britain's native plants. How can the scientists stop the seeds from germinating?

2/5

- **Keep the seeds dry.**
- **Keep the seeds in the dark.**
- **Keep the seeds cold.**
- **Keep the seeds warm.**
- **Keep the seeds damp.**

Plant Life Cycles Vocabulary

Objective: I can spell and define scientific words for plant life cycles.

Quick Search Ref: 10565

Level 1: Knowledge: I can spell and define scientific words for plant parts.

Required: 10/10 | Pupil Navigation: on | Randomised: off

1. _ _ _ _ _ _ is taken from a stamen to a stigma during pollination.

abc

- **Pollen**
- Poln
- Polun
- Pollun
- Polln
- Polen

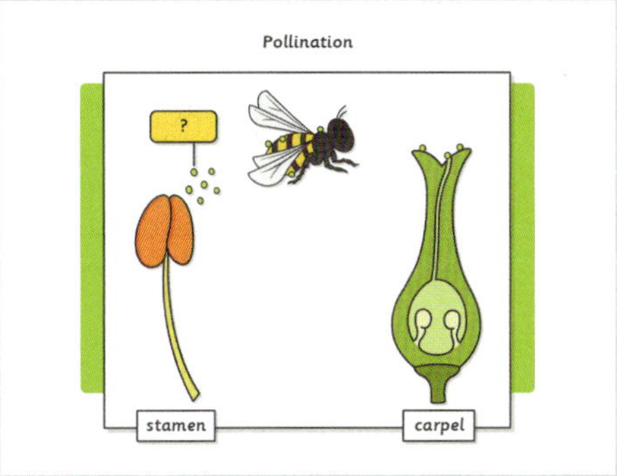

2. Pollen is transferred to the stigma during . . .

1/3

- fertilisation
- germination
- **pollination**

3. The male part of the flower is called the _ _ _ _ _ _.

abc

- staymun
- staymen.
- **stamen.**
- **stamen**
- staymun.
- staymen

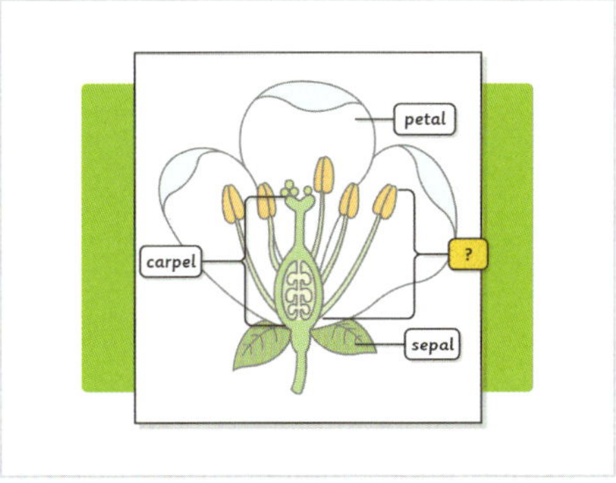

4. The sepals . . .

1/4

- produce pollen.
- attract insects to the flower.
- **protect the developing flower bud.**
- produce ovules.

5. Germination is when . . .

1/3

- **a seed grows a shoot and a root.**
- a new seed is made.
- a seed is spread from the parent plant.

6. Which part of the plant contains the ovules and develops into the fruit?

1/3

- the petals
- **the ovary**
- the stamen

Level 1 *continued*

7. The female part of the flower is called the _ _ _ _ _ _.

a b c

- pistal. ■ carpel ■ carpel. ■ pistil. ■ carpal
- pistil ■ pistal ■ pistol ■ carpul. ■ carpal.
- pistol. ■ carpul

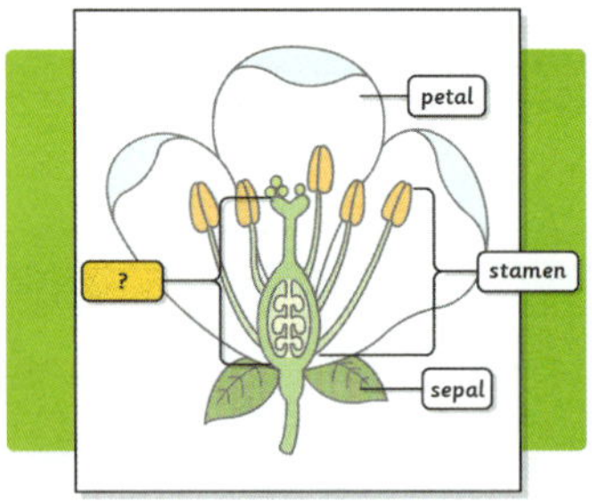

8. Seed ________ is how plants spread their seeds.

1/3

- fertilisation ■ germination ■ dispersal

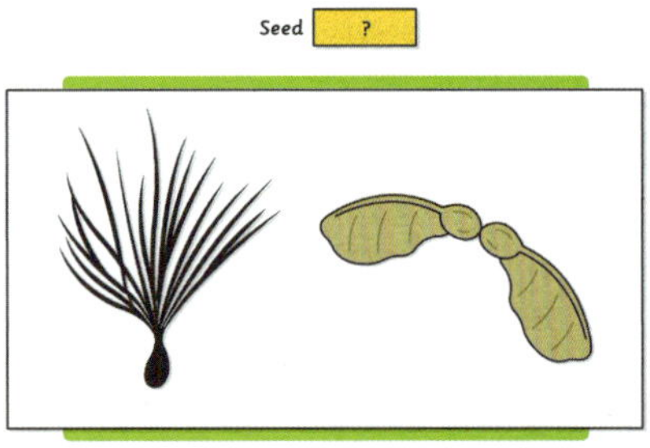

9. The _ _ _ _ _ _ is the top part of the carpel where the pollen lands.

a b c

- pistil ■ stigma ■ stigmar ■ carpel ■ stig

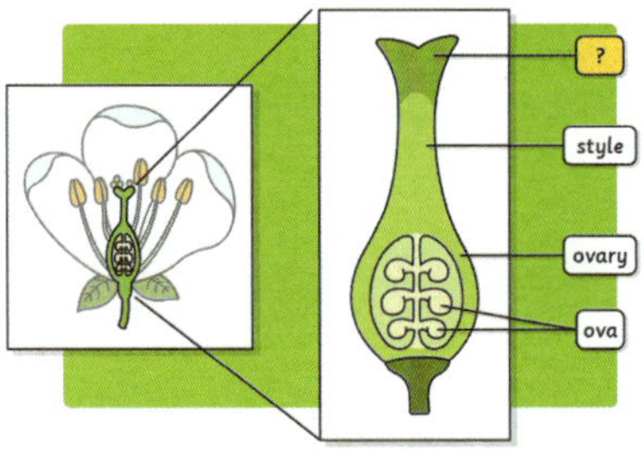

10. A _ _ _ _ is formed in the flower and can grow into a new plant.

a b c

- seyd ■ seed ■ sead ■ sede ■ sed

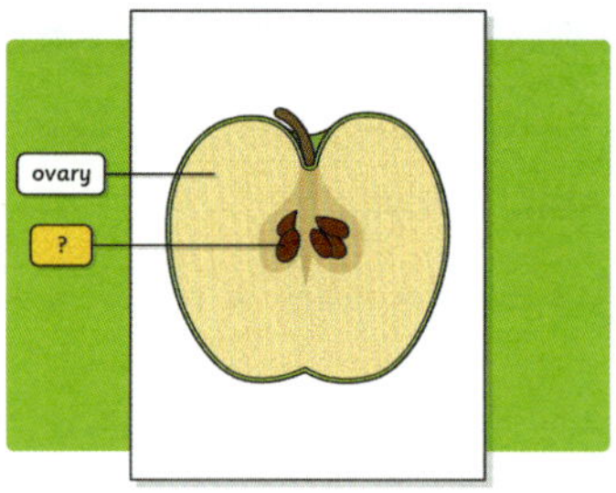

Plant Nutrition Vocabulary

Objective: I can spell and define scientific words for plant nutrition.

Quick Search Ref: 10512

Level 1: Knowledge: I can spell and define scientific words for plant nutrition.

Required: 10/10 | **Pupil Navigation:** on | **Randomised:** off

1. Plants make their own food. True or false?

1/2

- True
- False

2. Which part of the plant takes up water from the soil?

abc

- root
- roots
- the roots
- the routes
- the root
- routes
- route
- the route

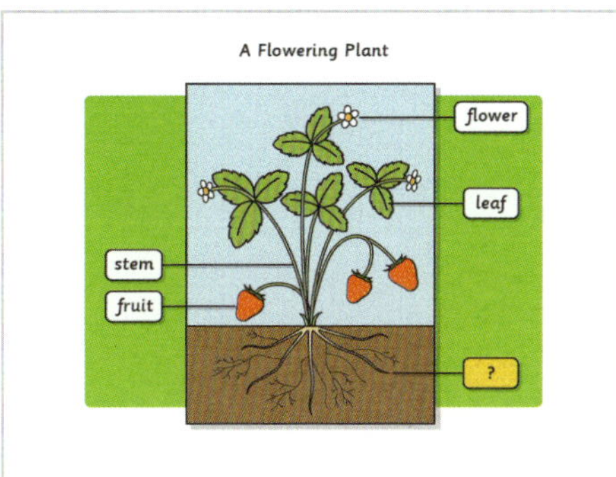

3. The stem holds the plant upright and . . .

1/4

- makes seeds.
- moves water around the plant.
- attracts insects.
- anchors the plant.

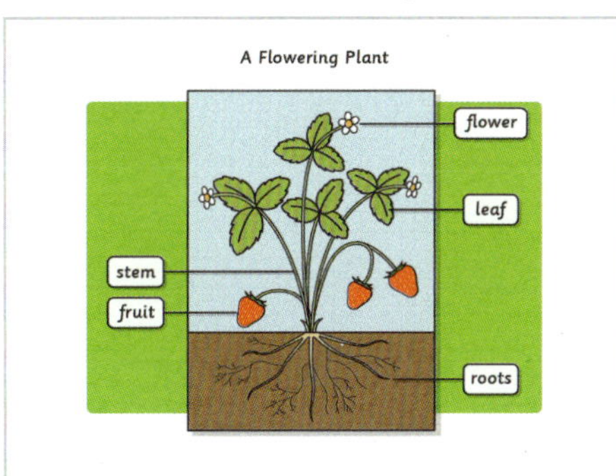

4. Green plants make their own food in their ______ by photosynthesis.

abc

- leaves
- leafs
- leyfs
- leefs

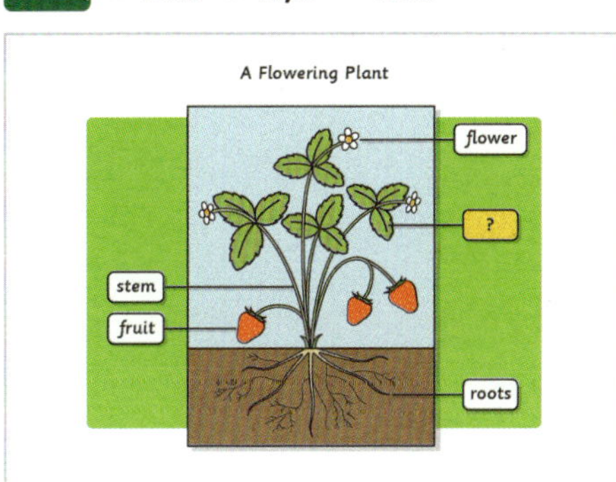

5. _____ in the leaves transport water and nutrients.

1/4

- Arteries
- Stems
- Veins
- Roots

6. Plants get water and _ _ _ _ _ _ _ _ _ through their roots.

abc

- newtrients
- nutrients
- newtriants
- nutriants

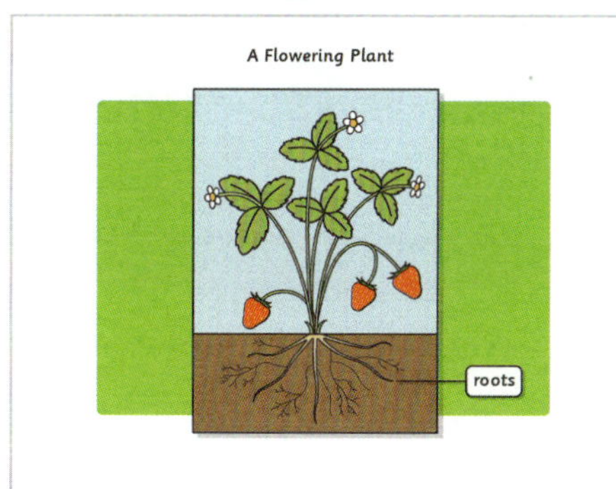

7. Which **2** parts of the plant are involved in water transportation?

2/4

- flower
- stem
- leaves
- roots

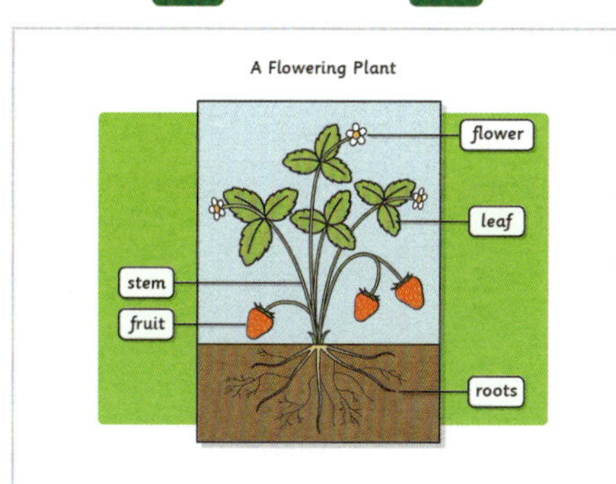

Level 1 *continued*

8. Farmers sometimes add fertiliser to the soil . . .

1/3

- to give plants more water.
- to help plants photosynthesise quicker.
- to give plants more nutrients.

Farmer Adding Fertiliser to the Soil

9. Photosynthesis is how green plants use energy from sunlight and water to . . .

1/4

- make seeds.
- make food.
- make flowers.
- make fruit.

A Flowering Plant

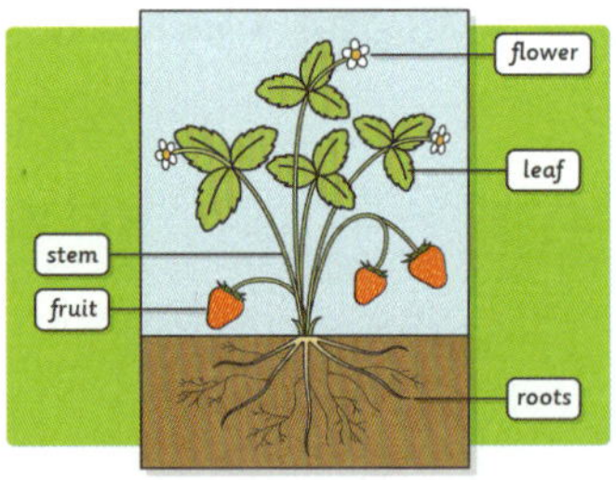

10. Which **3** of these are needed for plants to grow?

3/5

- Food made in the leaves.
- Pollen brought from other plants by insects.
- Nutrients taken up from the soil.
- Water taken up through the roots.
- Seeds being dispersed by wind.

A Flowering Plant

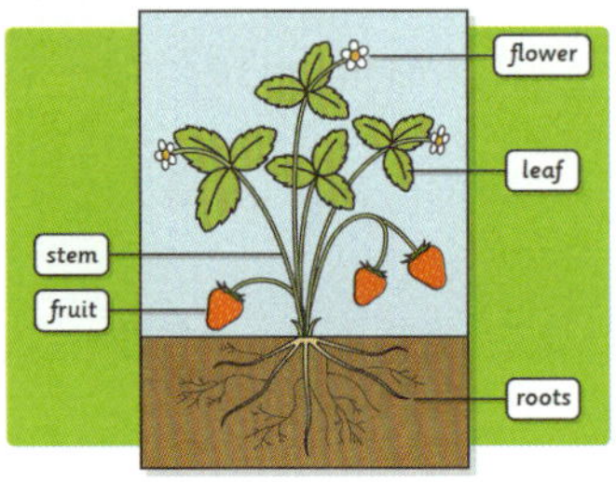

Primary Science Y3/4

Animals

Human Body
Movement
Nutrition

Human Skeleton, Muscles and Joints

Objective: I know that humans and some other animals have skeletons and muscles for support, protection and movement.

Quick Search Ref: 10605

Level 1: Knowledge: I can identify parts of the skeletal and muscular systems of a human.

Required: 8/8 **Pupil Navigation:** on **Randomised:** off

1. Which diagram shows the skeleton of a human?

1/4

- A
- B
- **C**
- D

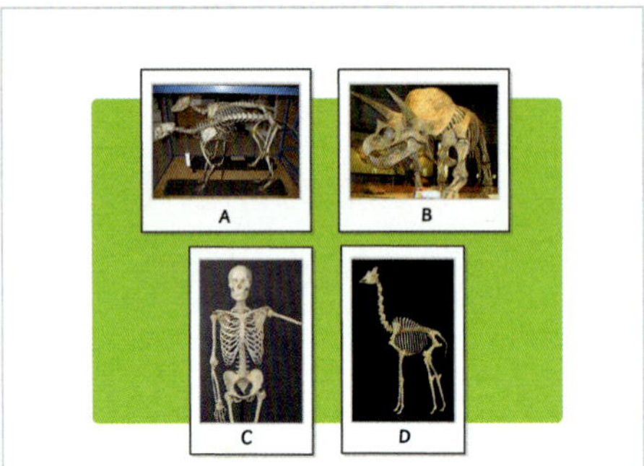

Triceratops Skeleton by Tyrell Museum

2. A skeleton is made up of _ _ _ _ _.

- **bones.**
- **bones**
- bowns

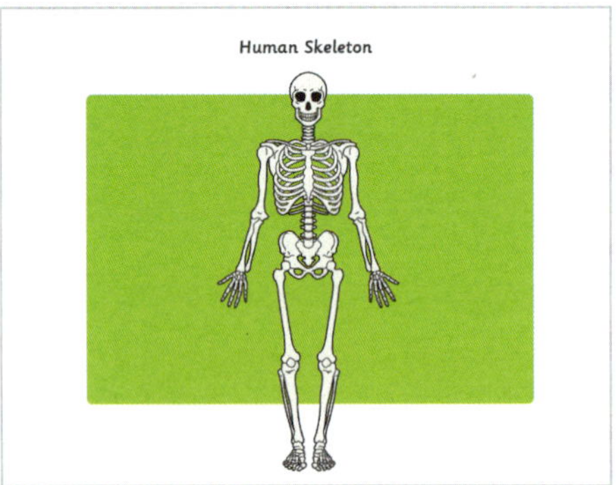

3. What does label **B** point to?

- **skull**
- scull
- scul
- head
- hed
- skul

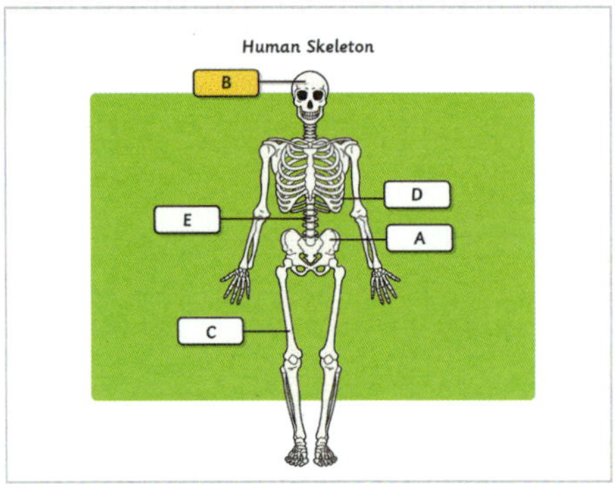

4. Which label points to the rib cage?

1/5

- A
- B
- C
- **D**
- E

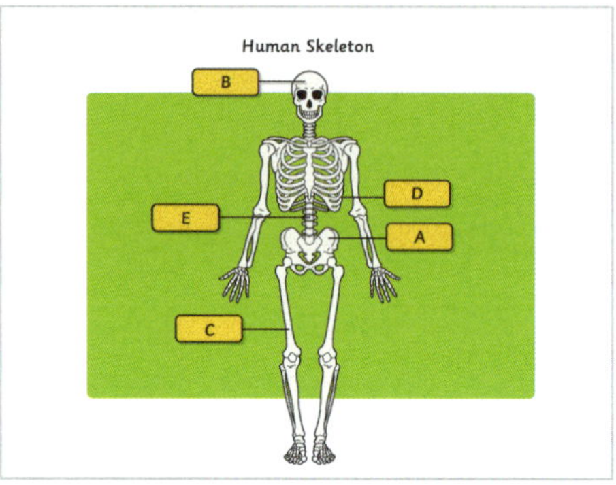

5. Which part of the human skeleton does the label point to?

1/3

- **spine**
- skull
- pelvis

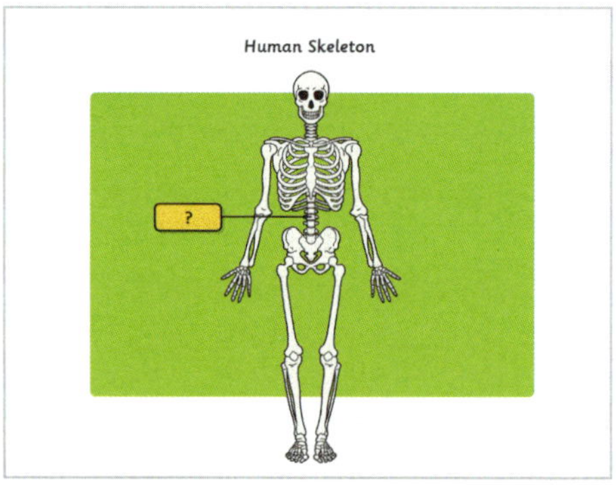

6. We have _ _ _ _ _ _ to help us bend our arms and legs.

- **joints**
- **joint**
- goints
- joynts
- joyns
- joins

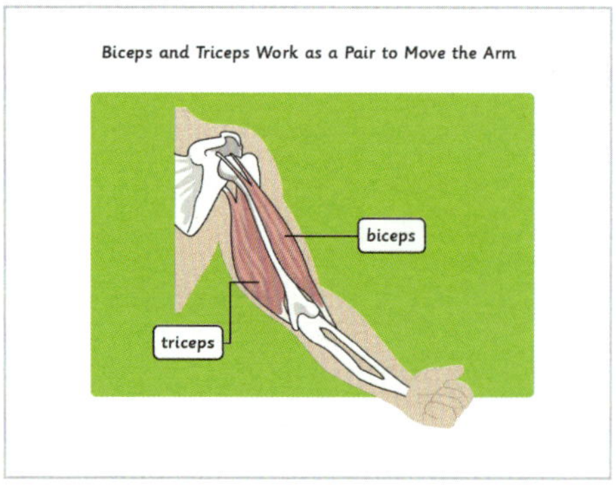

Level 1 *continued*

7. Which muscle does label **C** point to?

1/5

- radius
- biceps
- triceps
- ulna
- humerus

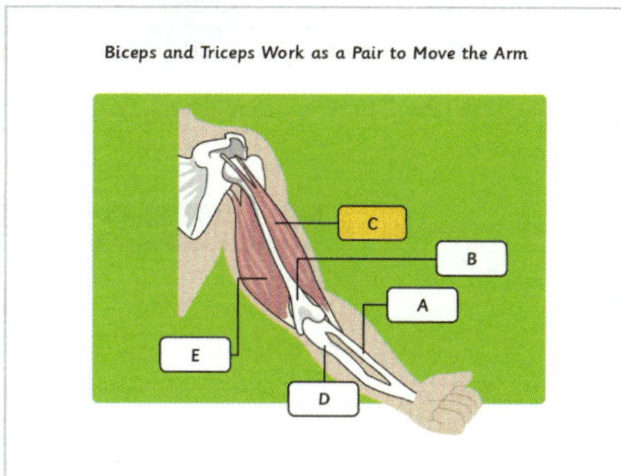

8. Which muscle does the label point to?

1/4

- biceps
- hamstring
- quadriceps
- triceps

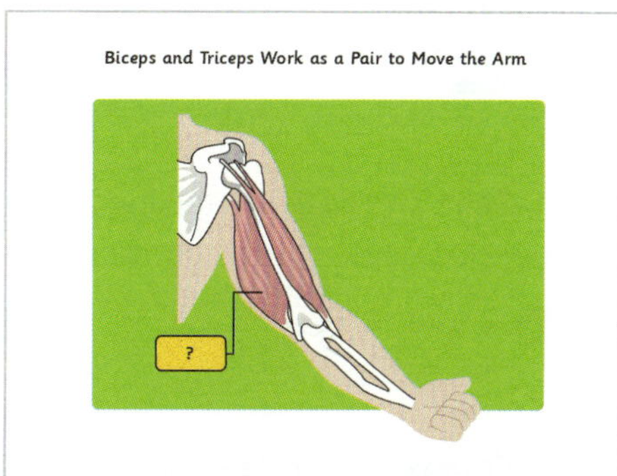

10. Your muscles help your skeleton to . . .

1/3

- protect your organs.
- move your body.
- support your body.

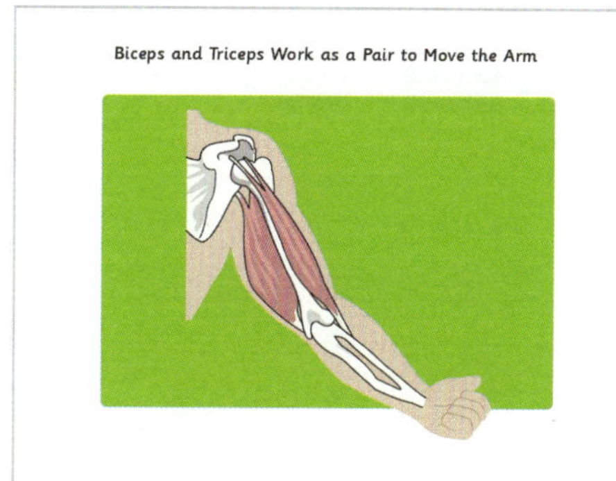

11. Muscles have to work in _ _ _ _ _ because they can only pull.

- pares
- pairs
- pears
- pers

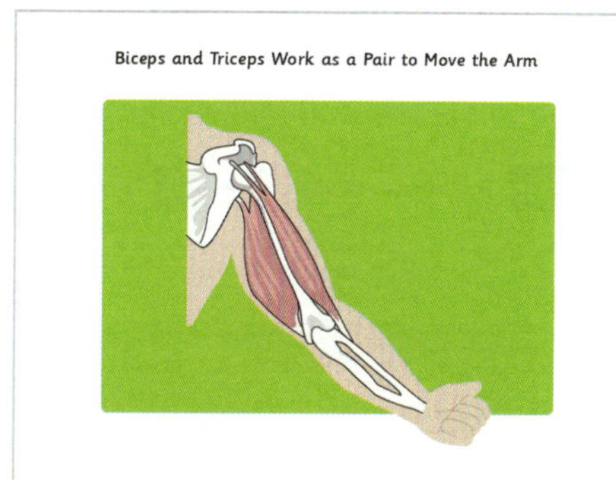

Level 2: Understanding: I can explain how the muscular and skeletal systems work together to create movement.

Required: 7/7 **Pupil Navigation:** on

Randomised: off

9. What are the three jobs of the human skeleton?

3/5

- to turn food into energy
- to protect your organs
- to help you to move
- to help blood flow around the body
- to support your body

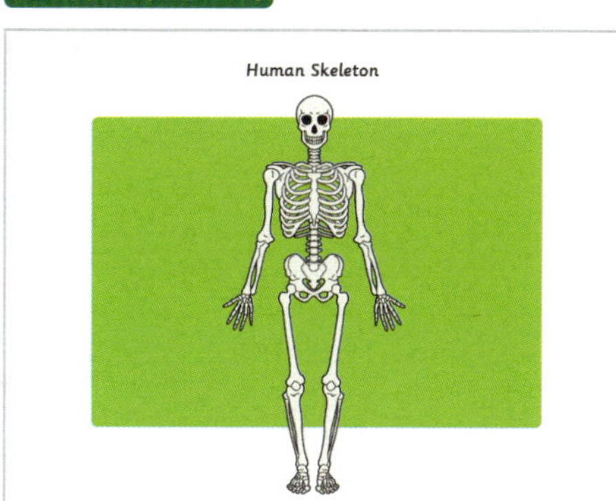

12. When a muscle contracts, it gets shorter. Which muscle gets shorter when you bend your arm?

- bicep
- biceps
- tricep
- bicept
- bicepts
- triceps

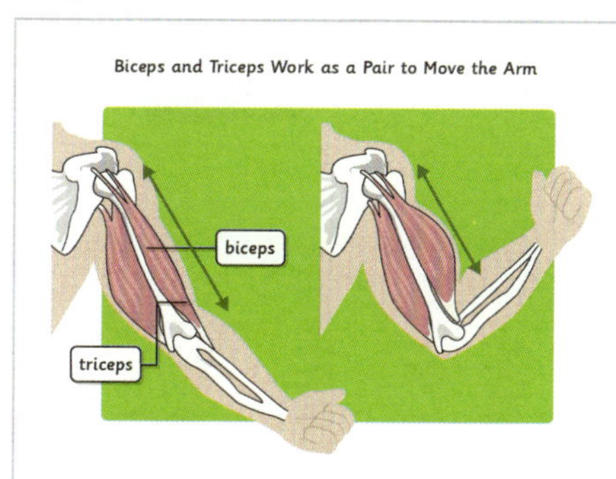

Level 2 *continued*

13. When you straighten your arm, your triceps muscle contracts and your biceps muscle _ _ _ _ _ _ _.

a b c

- relacs
- **relaxes**
- **relax**
- relaxis
- relacis
- relacses

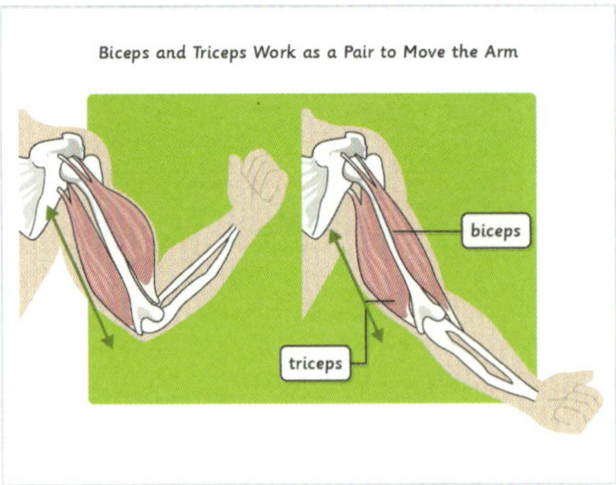

14. Ball and socket joints help you to swing your arms and legs in different directions. Which of these joints is a ball and socket joint?

1/5

- **shoulder**
- elbow
- knee
- **hip**
- neck

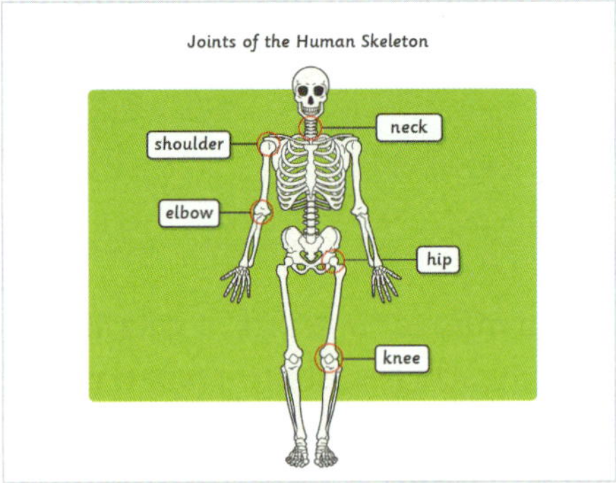

15. Knee and elbow joints move in a similar way to a door with hinges. They are . . .

1/3

- ball and socket joints.
- **hinge joints.**
- pivot joints.

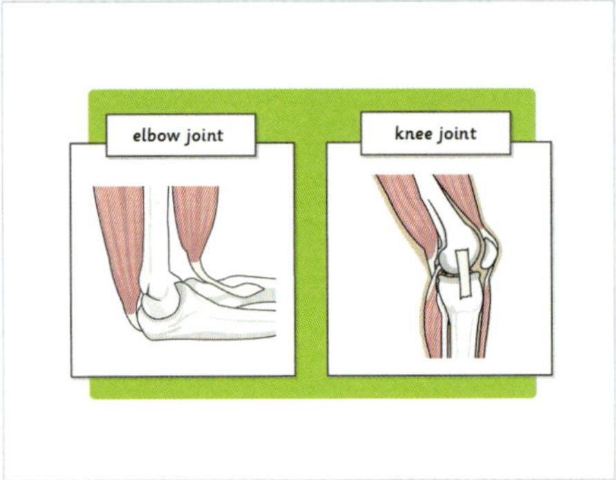

Level 3: Application: I can apply my knowledge and understanding of skeletons, muscles and joints to the world around me.

Required: 3/3 **Pupil Navigation:** on
Randomised: off

16. Which muscles contracts when Rosie bends her knee?

1/4

- quadriceps muscles
- **hamstring muscles**
- shin muscles
- calf muscles

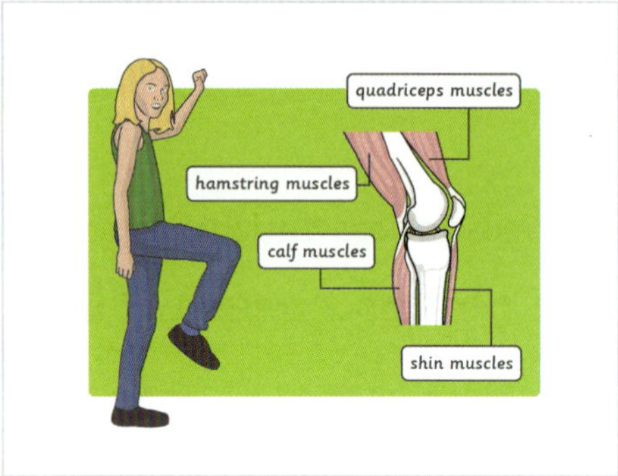

17. Naresh, Bella and Max are talking about how their muscles work. Who is correct?

1/3

- **Naresh**
- Bella
- Max

18. When Jonah broke his right leg, he had a cast on it for 2 months. When his cast was taken off, his right leg looked thinner than his left leg. Why do you think this is?

a b c

- Open question, no set answer

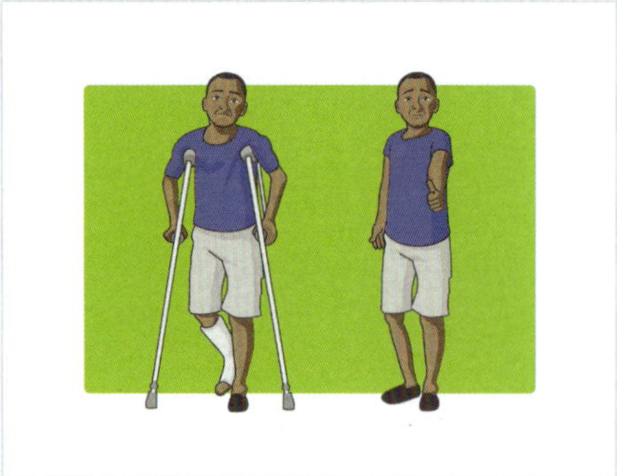

Skeletons, Muscles and Joints Vocabulary

Objective: I can use vocabulary relating to skeletons, muscles and joints.

Quick Search Ref: 10596

Level 1: Knowledge: I can spell and define scientific words for skeletons, muscles and joints.

Required: 10/10 **Pupil Navigation:** on **Randomised:** off

1. Your _ _ _ _ _ _ _ _ supports and protects your body. It also helps you to move.

abc

- skelington
- **skeleton**
- sceliton
- sceleton
- skeliton

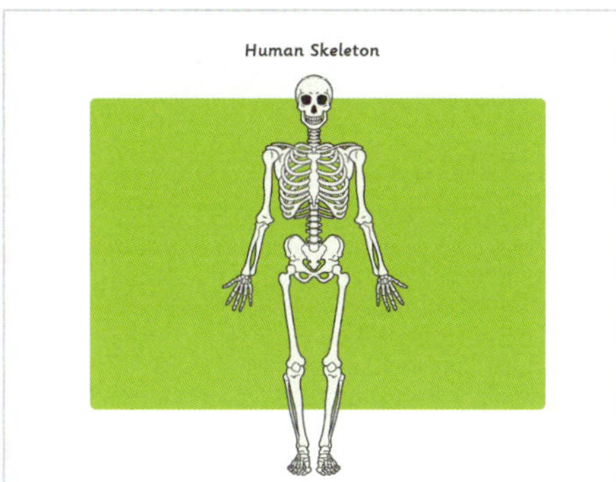

2. ___________ is an organism with an internal skeleton.

1/2

- An invertebrate
- **A vertebrate**

giraffe, cat by Philip Hartwell; crocodile by Tomascastelazo

3. The _ _ _ _ _ bones protect your brain.

abc

- **skull**
- **your skull**
- **the skull**
- **skulls**
- scull
- **a skull**
- scul
- skul

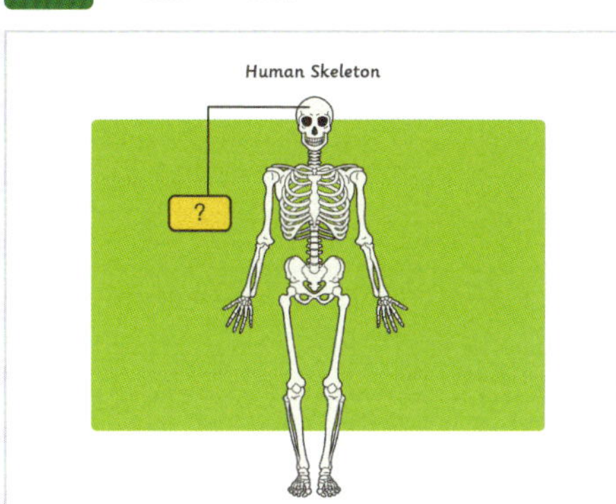

4. Your _ _ _ cage protects your lungs and heart.

abc

- **rib**

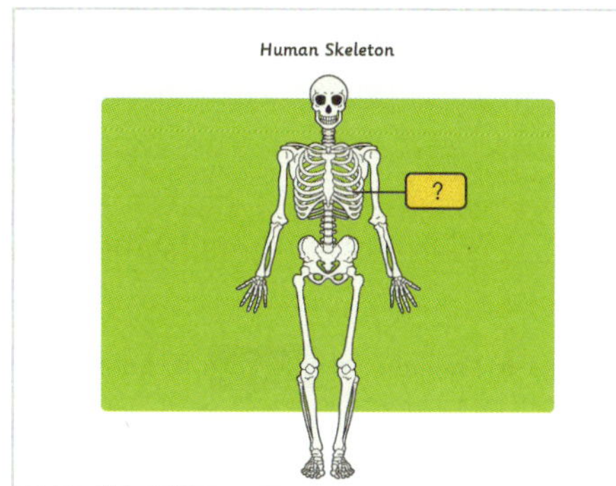

5. An invertebrate is . . .

1/3

- an organism with an internal skeleton.
- an organism with an exoskeleton.
- **an organism that doesn't have an internal skeleton.**

butterfly by Francesca Hartwell

6. Some invertebrates have a skeleton on the outside of their bodies. This is called an _ _ _ skeleton.

abc

- **exo**
- **exoskeleton**
- xoskeleton
- xo

snail by Philip Hartwell; dragonfly nymph by Greg Schechter

Level 1 *continued*

7. A _ _ _ _ _ is where two bones meet.

a b c

- joint
- goint
- jont
- joynt

A Hinge Joint

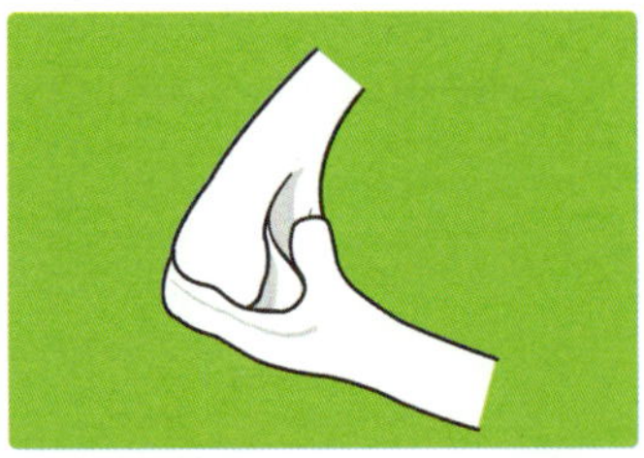

8. _________ relax and contract to move our bones.

1/4

- Organs
- Skeleton
- Muscles
- Joints

Muscles of the Arm Joint

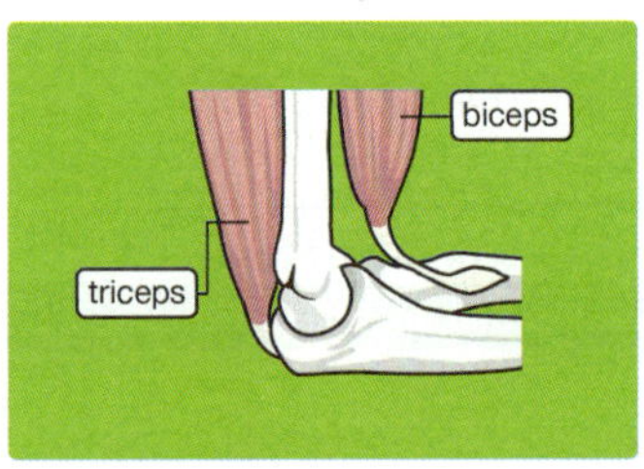

9. A __________ muscle gets shorter and pulls on the bone to move it.

1/2

- contracting
- relaxing

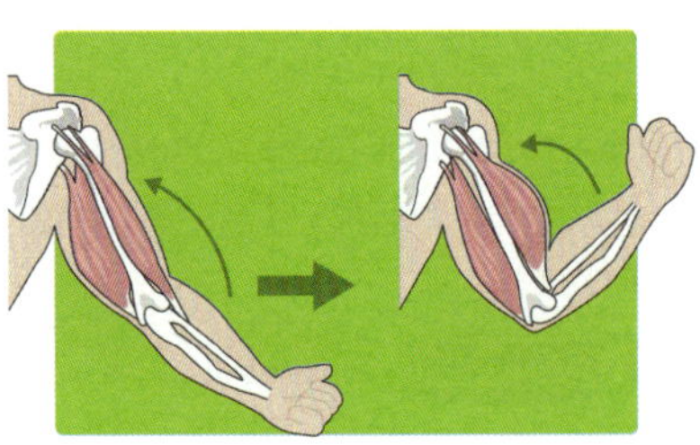

10. When muscles _ _ _ _ _, they go back to their normal size and stop pulling on the bone.

a b c

- relax
- relacs

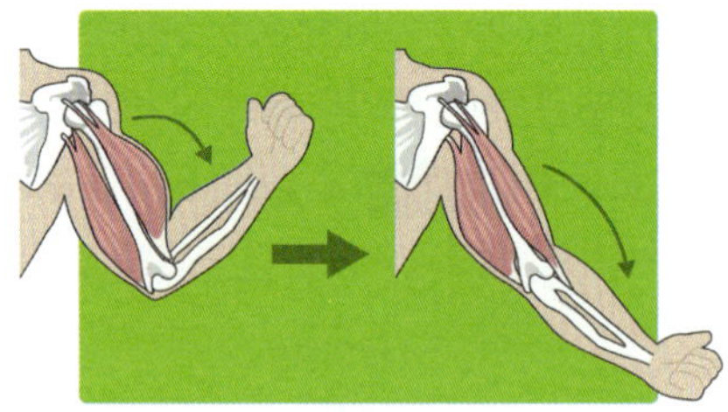

Vertebrates and Invertebrates (Including Exoskeletons)

Objective: I know that humans and some other animals have skeletons for support, protection and movement.

Quick Search Ref: 11017

Level 1: Knowledge: I can identify vertebrates and invertebrates.

Required: 8/11 **Pupil Navigation:** on **Randomised:** off

1. Scientists group living things based on their features. The animals in the diagram have been grouped based on whether or not they . . .

1/2

- have skeletons inside their bodies.
- are animals or insects.

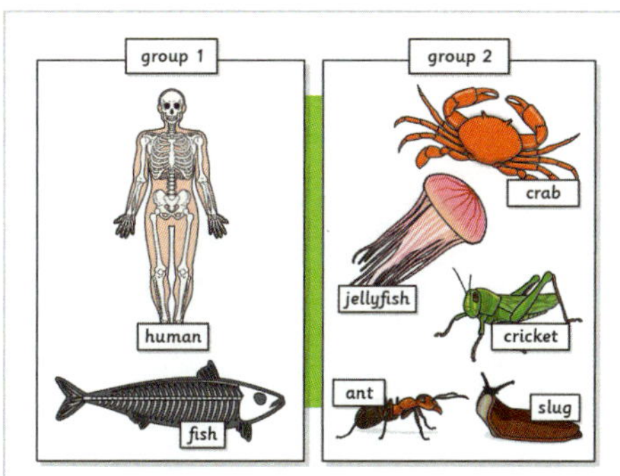

2. Animals that have skeletons inside their bodies (internal skeletons) are called __________.

1/2

- vertebrates
- invertebrates

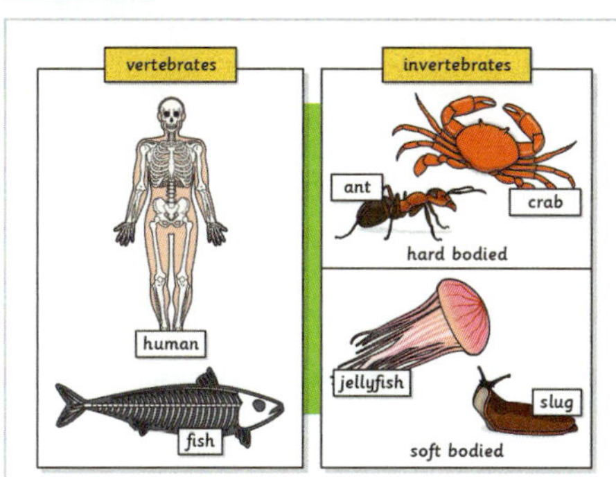

3. Animals that don't have internal skeletons are called _________.

1/2

- vertebrates
- invertebrates

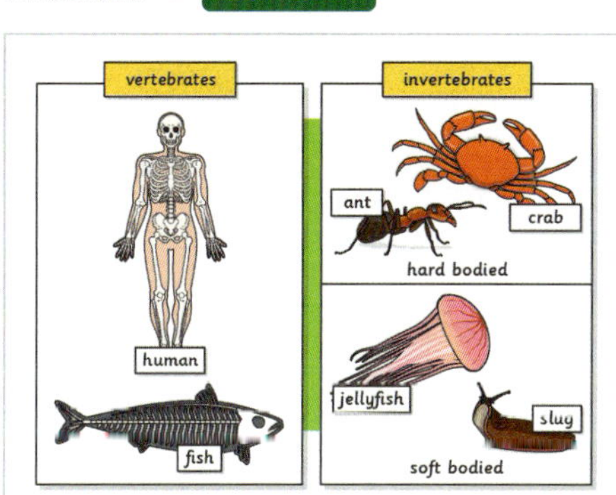

4. Is a human a vertebrate or invertebrate?

1/3

- an invertebrate
- a vertebrate
- neither because a human isn't an animal

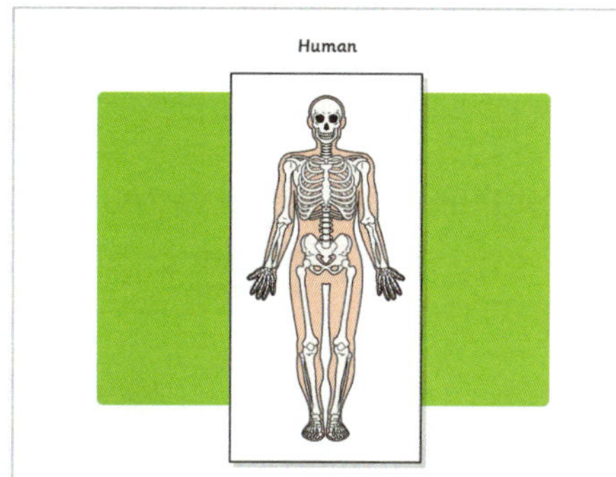

5. Is a slug a vertebrate or an invertebrate?

1/2

- an invertebrate
- a vertebrate

slug by Francesca Hartwell

6. Is a giraffe a vertebrate or an invertebrate?

1/2

- an invertebrate
- a vertebrate

giraffe by Phillip Hartwell

Level 1 *continued*

7. Which of these animals are invertebrates?

2/4

- **ant**
- **spider**
- parrot
- horse

spider by Francesca Hartwell; *horse*, *parrot* by Phillip Hartwell; *ant* by Rachel Hammerton

8. Which of these animals are vertebrates?

2/5

- **koala**
- caterpillar
- **elephant**
- ladybird
- butterfly

koala, *ladybird*, *butterfly* by Francesca Hartwell; *elephant* by Phillip Hartwell; *caterpillar* by Gail Hampshire

9. Is a frog a vertebrate or an invertebrate?

1/2

- **a vertebrate**
- an invertebrate

frog by H Krisp

10. Which of these animals are invertebrates?

3/5

- **greenfly**
- sheep
- **cricket**
- **caterpillar**
- duck

greenfly by Francesca Hartwell; *sheep*, *duck*, *cricket* by Phillip Hartwell; *caterpillar* by Gail Hampshire

11. A hedgehog is . . .

1/2

- an invertebrate.
- **a vertebrate.**

hedgehog by Phillip Hartwell

Level 2: Understanding: I understand that invertebrates can have soft or hard bodies.

Required: 6/7 **Pupil Navigation:** on

Randomised: off

12. Animals can be classified as either vertebrates or invertebrates. What is an invertebrate?

1/4

- an organism with a hard covering on the outside of their body.
- **an animal without an internal skeleton**
- an insect
- an animal with an internal skeleton

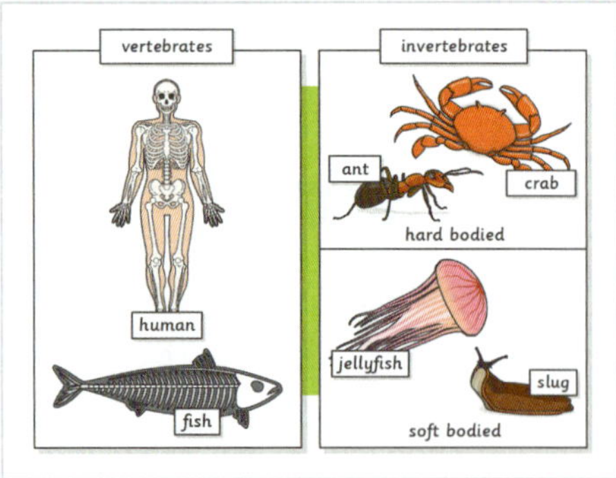

Level 2 *continued*

13. Invertebrates can have hard or soft bodies. Which **two** animals are hard-bodied invertebrates?

2/4

- cow
- slug
- ant
- crab

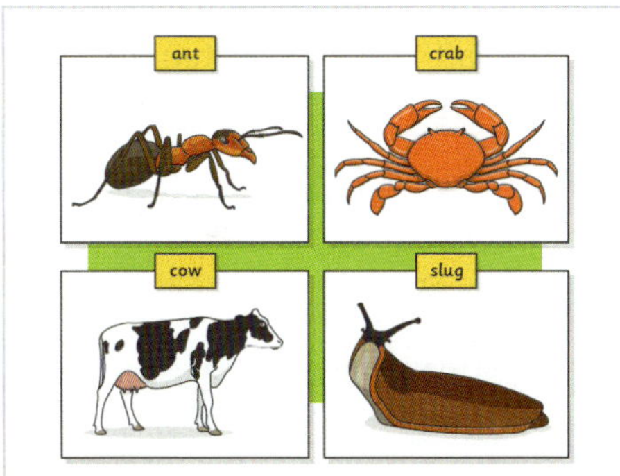

14. Invertebrates can have hard or soft bodies. Select the animals that are invertebrates that have soft bodies.

2/5

- slug
- spider
- scorpion
- worm
- snake

slug by Francesca Hartwell; *worm* by Dodo Bird; *cockroach* by Ian Sutton

15. The animals in the diagram all have an exoskeleton. What is an exoskeleton?

1/2

- a hard protective layer that covers the body of an animal.
- a skeleton on the inside of the body

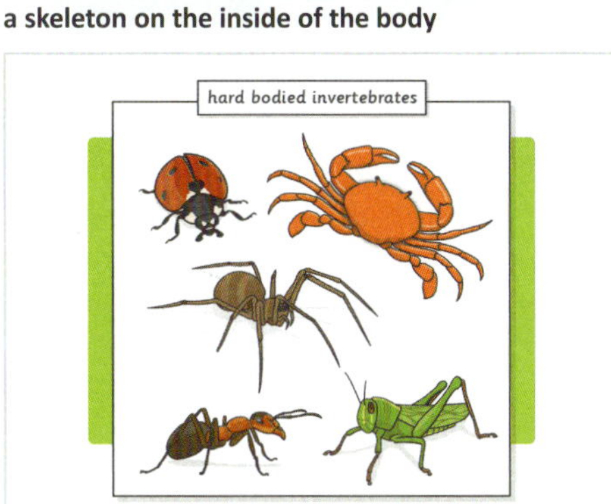

16. Which **three** of these animals have exoskeletons?

3/5

- bee
- slug
- worm
- cricket
- crab

17. The scientific name for the hard outer covering of a hard-bodied invertebrate is an exoskeleton. Select **three** invertebrates that have exoskeletons.

3/5

- slug
- spider
- scorpion
- worm
- cockroach

slug by Francesca Hartwell; *worm* by Dodo Bird; *cockroach* by Ian Sutton

18. Invertebrates can have hard or soft bodies. Select the three invertebrates that have soft bodies.

3/5

- jellyfish
- slug
- ant
- ladybird
- worm

Level 3: Application: I can apply my knowledge about vertebrates and invertebrates.

Required: 3/3 **Pupil Navigation:** on
Randomised: off

19. The picture shows a snake skeleton. Is a snake a vertebrate or an invertebrate? Explain your answer.

- Open question, no set answer

Snake Skeleton

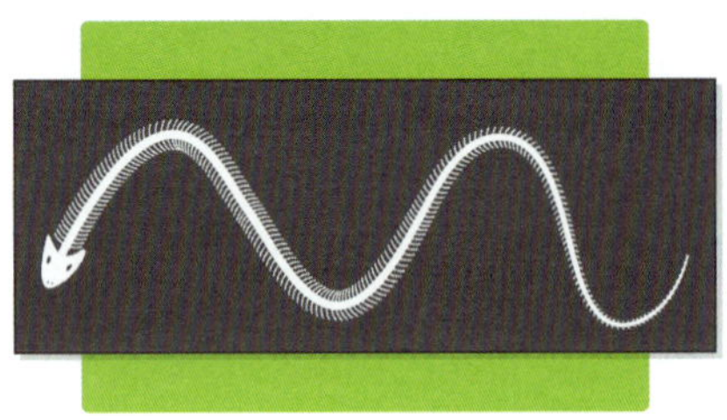

20. Jason is sorting his collection of animal cards into two piles: invertebrates and vertebrates. Which fact is he using to classify the animals?

1/4

- skin covering
- size
- **backbone or no backbone**
- diet

21. A tortoise has a hard outer covering and a backbone. Is a tortoise a vertebrate or an invertebrate? Explain your answer.

- Open question, no set answer

Tortoise

Level 4: Extended reasoning: I can apply my knowledge and understanding of vertebrates and invertebrates to new contexts.

Required: 1/2 **Pupil Navigation:** off
Randomised: on

22. Ethan and his friends are discussing whether a bee is a soft-bodied invertebrate or a hard-bodied invertebrate. Who is correct?

1/3

- Ethan
- Grace
- **Irma**

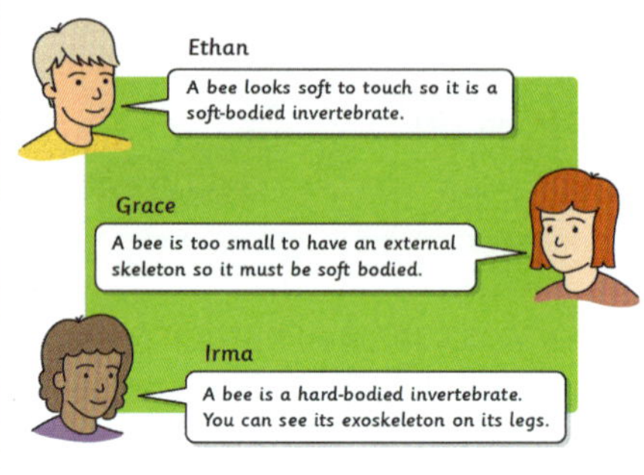

23. An animal with an exoskeleton grows by shedding its exoskeleton. Its new exoskeleton is soft at first but then hardens to form a protective covering. Animals often hide away when they are shedding their exoskeletons. Why do you think this is?

- Open question, no set answer

Cricket

Animal Nutrition Vocabulary

Objective:	I can use vocabulary relating to animal nutrition.
Quick Search Ref:	10711

Level 1: Knowledge: I can spell and define scientific words for animal nutrition.

Required: 11/11 **Pupil Navigation:** on **Randomised:** off

1. A __________ eats mainly meat.

a b c

- carnyvor
- carnivore
- herbivore
- carnivores
- omnivore
- carniyvor
- carnivor

lion by Phillip Hartwell; *great diving beetle* by Evanherk; *shark* by Tanaka Juuyoh

2. Omnivores . . .

1/3

- eat mainly meat.
- eat meat and plants.
- eat mainly plants.

badger by Kokay Seabolcs

3. What do herbivores mainly eat?

a b c

- plants

horse & sheep by Phillip Harwell

4. An animal's ____ is what it eats.

a b c

- diet
- diat
- diate
- diyet

5. Sugary foods such as sweets, and starchy foods such as potatoes are sources of . . .

1/4

- proteins.
- fats.
- vitamins and minerals.
- carbohydrates.

6. We need _______ to help our bodies grow and repair.

a b c

- protein
- proteins
- proteyn
- prowteen
- proteen
- protien

7. ______ in our diet helps us to stay warm.

1/3

- Carbohydrates
- Proteins
- Fats

Level 1 *continued*

8. It is important to eat lots of fruit and vegetables to get different _ _ _ _ _ _ _ _ _ and minerals.

a b c

- vitamins
- vitamin

9. Eating lots of _ _ _ _ _ is important for healthy digestion.

a b c

- fibar
- fibre
- fybre
- fybar
- fiber
- fyber

10. Having a healthy balanced diet means eating the right amount of each food type and drinking enough _ _ _ _ _ .

a b c

- warter
- water.
- water
- watar
- watr
- woter
- watr.
- warter.
- watar.
- woter.

11. _ _ _ _ _ _ _ _ is a mineral found in milk and cheese that helps our bones stay strong and healthy.

a b c

- Calcium
- Calseeum
- Calcum
- Calsum
- Calsium

Different Animal Diets

Objective: I know that humans and other animals need specific types and amounts of nutrition.

Quick Search Ref: 10560

Level 1: Knowledge: I can identify different diets and the animals that have them.

Required: 7/9 **Pupil Navigation:** on **Randomised:** off

1. A carnivore . . .

1/3

- eats meat and plants.
- mainly eats plants.
- **eats mainly meat.**

lion by Phillip Hartwell; *great diving beetle* by Evanherk; *shark* by Tanaka Juuyoh

2. A _ _ _ _ _ _ _ _ _ _ eats mainly plants.

- herbevor
- herbvor
- **herbivore**
- herbvore
- herbivor
- herbevore

horse & sheep by Phillip Harwell

3. What do omnivores eat?

1/3

- mainly plants
- mainly meat
- **meat and plants**

badger by Kokay Seabolcs

4. Which of these animals are carnivores?

2/7

- **cat**
- elephant
- giraffe
- horse
- gibbon
- **tiger**
- human

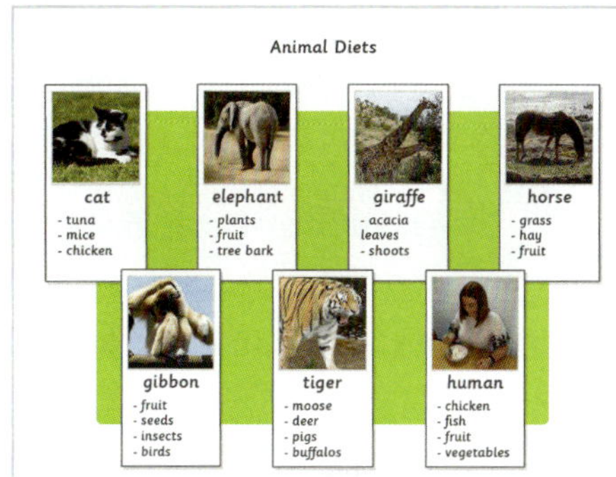

5. Which of these animals are herbivores?

3/7

- cat
- **elephant**
- **giraffe**
- **horse**
- gibbon
- tiger
- human

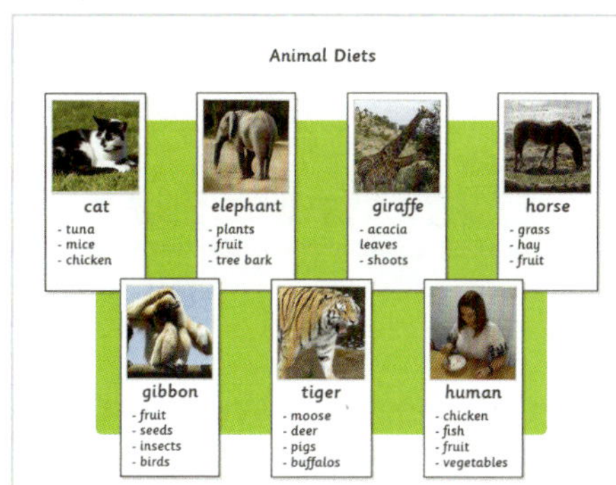

6. Dolphins are carnivores. Which of these foods would a dolphin eat?

2/5

- **mackrel**
- seaweed
- coral
- **herring**
- seagrass

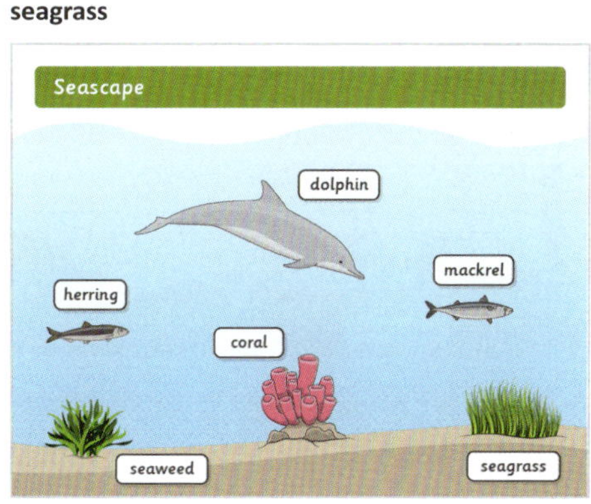

Level 1 *continued*

7. Macaws are herbivores. Which of these foods would a macaw eat?

3/6

- Brazil nut
- frog
- sloth
- butterfly
- leaves
- mango

8. Select the omnivores from this group of animals.

2/7

- cat
- elephant
- giraffe
- horse
- gibbon
- tiger
- human

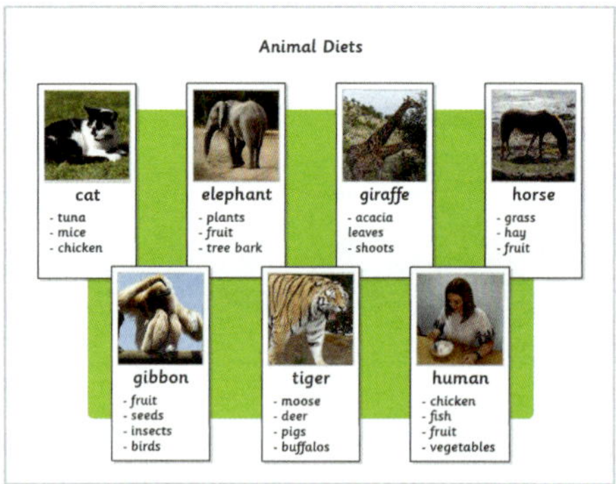

9. Badgers are omnivores. Which of these foods would a badger eat?

1/7

- blackberry
- apple
- beetle
- mushroom
- hedgehog
- frog
- all of these things

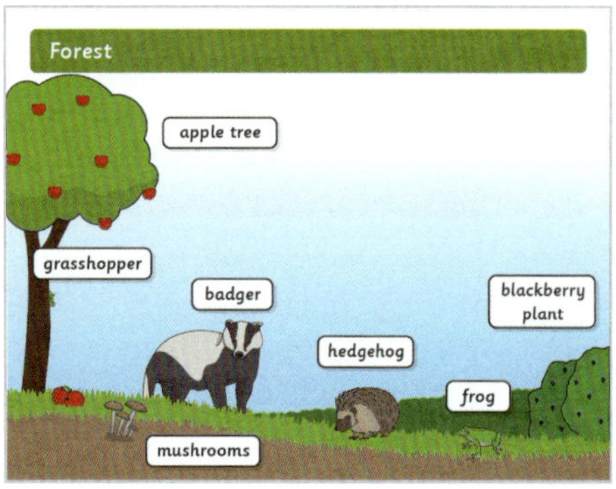

Level 2: Understanding: I can describe how animals get nutrients from their food.

Required: 5/5 **Pupil Navigation:** on
Randomised: off

10. Chloe, James and Amir are talking about diets. Who do you agree with?

1/3

- Chloe
- James
- Amir

11. Animals need to eat food because it gives them ______ to move and grow.

- enargie
- energie
- enargy
- energy
- enagie
- enegie
- enagy
- enegy

12. Select the statement that is **true** for all these animals.

1/4

- They all eat mainly meat.
- They don't make their food inside their own bodies.
- They are all herbivores.
- They all need the same amount of food.

Level 2 continued

13. Plants don't need to eat other living things because . . .

1/3

- they make their own food in their leaves.
- they don't need energy.
- they get nutrients from the soil.

14. A balanced human diet would include which of the following things?

1/6

- vitamins and minerals
- sugary carbohydrates
- protein
- starchy carbohydrates
- fats
- all of these things

Level 3: Application: I can apply my knowledge and understanding of different diets to the world around me.

Required: 3/3 **Pupil Navigation:** on
Randomised: off

15. Melissa, Sam and Milo are discussing different human diets. Who do you agree with?

1/3

- Melissa
- Sam
- Milo

16. In the habitat shown, the badger will eat fruit and small animals. If all of the hedgehogs were removed, what might happen to the badger's diet?

1/3

- The badger will have to eat more fruit to replace the hedgehogs.
- The badger won't have enough food, so it won't survive.
- The badger will have to replace hedgehogs with other small animals.

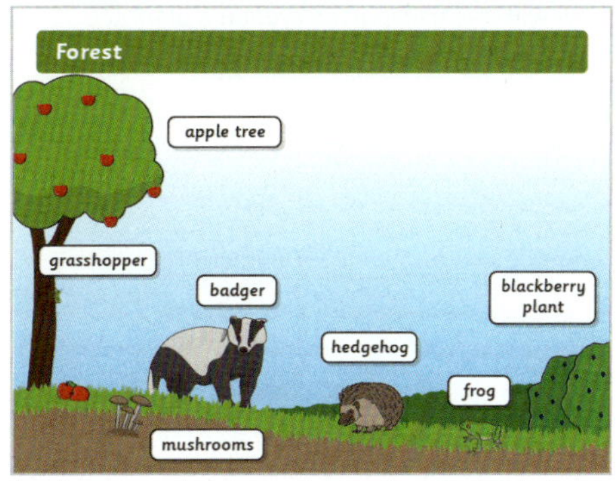

17. Dairy foods such as cheese and milk contain calcium, which we need to grow healthy bones and teeth. If you were allergic to dairy products, what **two foods** could you eat to make sure you get enough calcium?
Use the information cards to help you.

2/5

- rice
- green beans
- candy floss
- yoghurt
- tuna

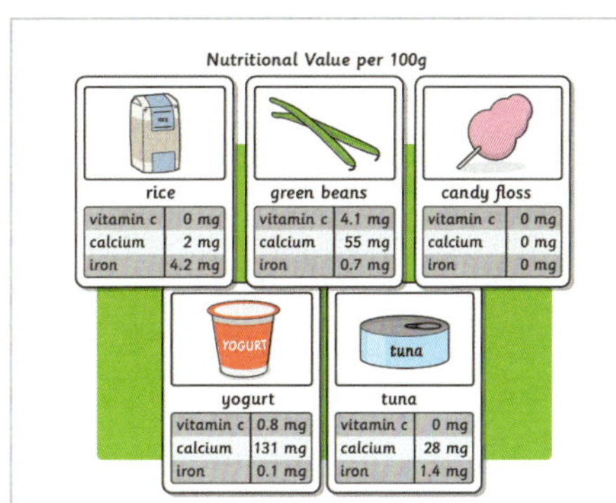

Digestion Vocabulary

Objective: I can use key words related to digestion.

Quick Search Ref: 10451

Level 1: Knowledge: I can spell and define scientific words for digestion.

Required: 14/14 **Pupil Navigation:** on **Randomised:** off

1. Digestion is . . .

- how blood is pumped around the body.
- how water is transported around a plant.
- **the process of breaking down food.**

1/3

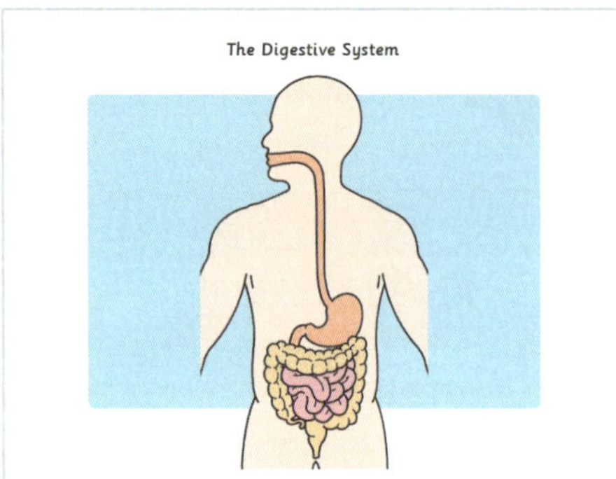

2. Digestion starts in the _ _ _ _ _.

- mowf
- **mouth.**
- **mouth**
- mowth.
- mowf.
- mowth
- mouf.
- mouf

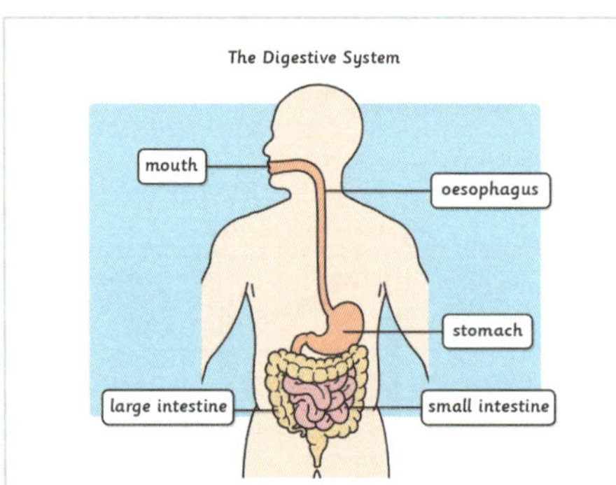

3. Food travels from the mouth to the stomach through the _____________.

- tongue
- windpipe
- **oesophagus**
- large intestine

1/4

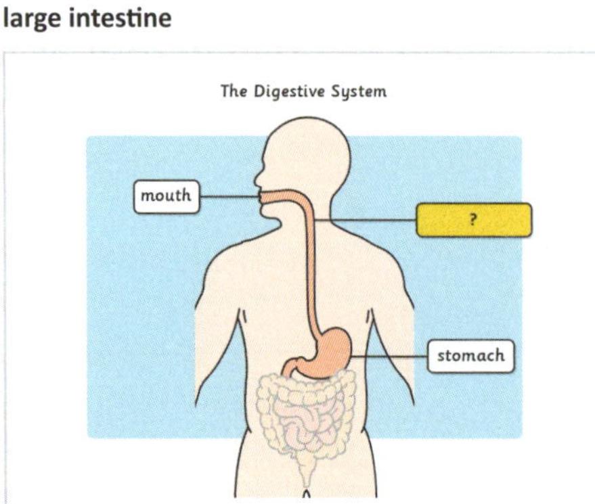

4. One function of the stomach is to . . .

- store food.
- **add acid to food in order to kill bacteria.**
- absorb nutrients from food.

1/3

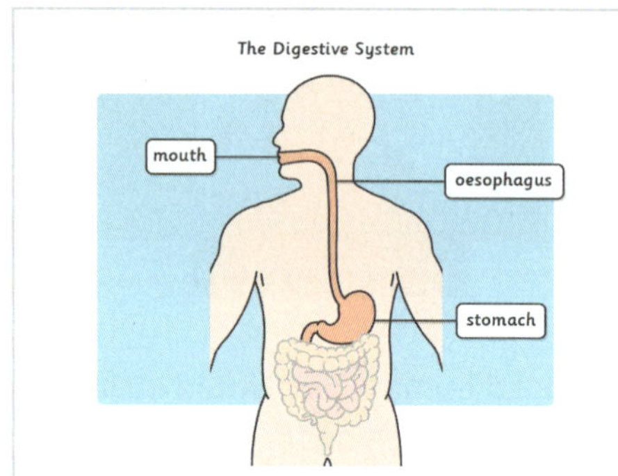

5. The small intestine absorbs _________ from food.

- water
- oxygen
- **nutrients**

1/3

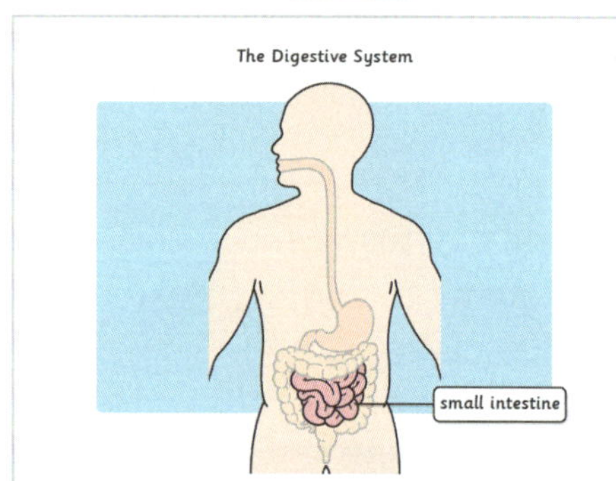

6. The large intestine . . .

- absorbs nutrients following digestion.
- **absorbs water following digestion.**
- absorbs oxygen during digestion.

1/3

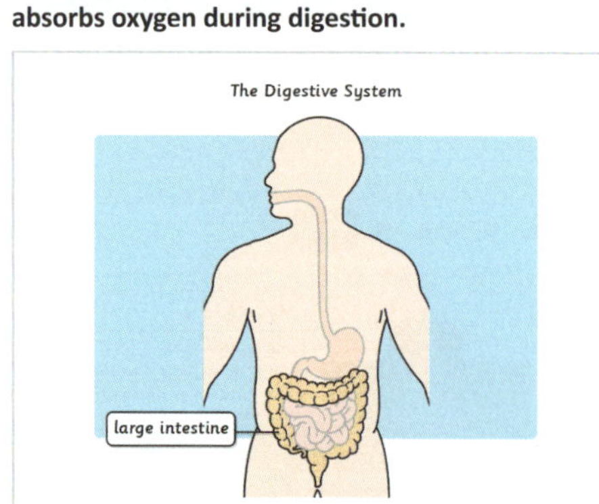

Level 1 *continued*

7. The anus is where . . .

☐ ☒ ☐

- **undigested waste leaves the body.**
- **food is broken down.**
- **where food enters the body.**

1/3

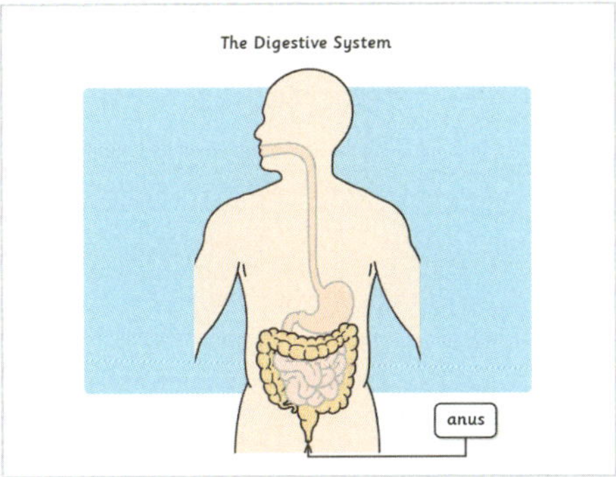

8. We get nutrients from our food, which help us live and _ _ _ _.

a b c

- **grow**
- **groe**
- **grow.**
- **gro.**
- **groe.**
- **gro**

9. _ _ _ _ _ _ is the hard outer part of the tooth.

a b c

- **Anamel**
- **Enamel**
- **Inamel**
- **Enamul**
- **Enamal**

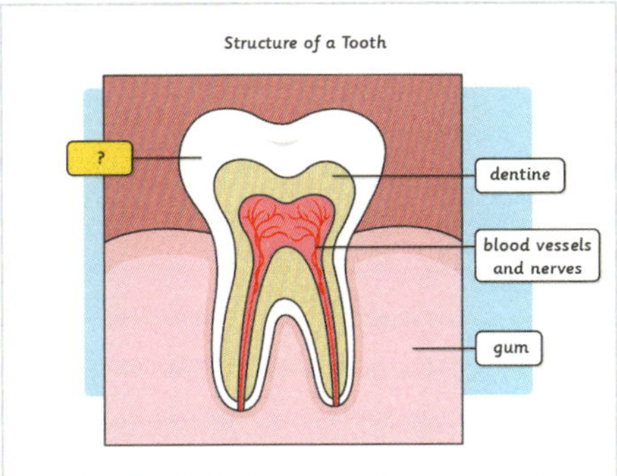

10. We use our molar teeth for . . .

☐ ☒ ☐

- **gripping, shredding and tearing food.**
- **grinding and chewing food.**
- **biting into food.**
- **crushing and tearing food.**

1/4

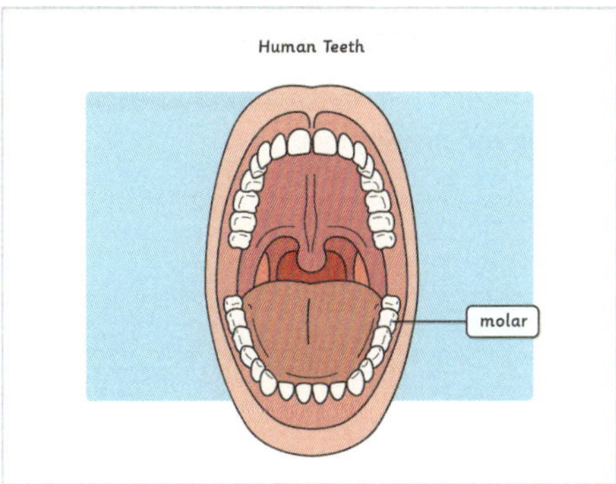

11. We use our _ _ _ _ _ _ teeth for gripping, shedding and tearing.

a b c

- **canine**
- **canin**
- **knine**
- **kanine**

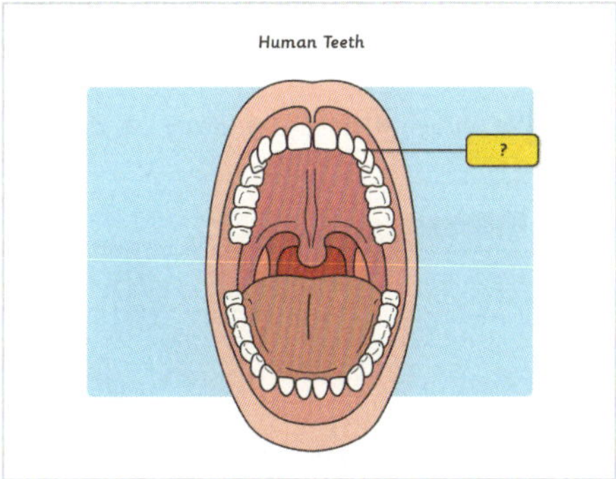

12. Which teeth do we use for biting into food?

☐ ☒ ☐

- **premolar**
- **canine**
- **molar**
- **incisor**

1/4

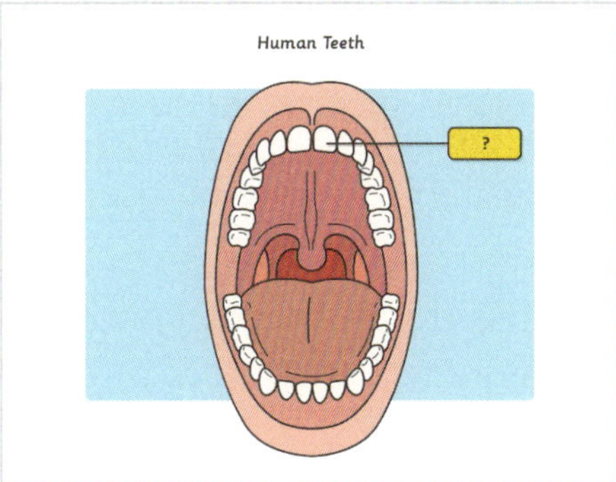

Level 1 *continued*

13. Our premolar teeth are between our canine teeth and our molars. We use our premolar teeth for . . .

1/4

- gripping, shredding and tearing food.
- biting into food.
- crushing and tearing food.
- grinding and chewing food.

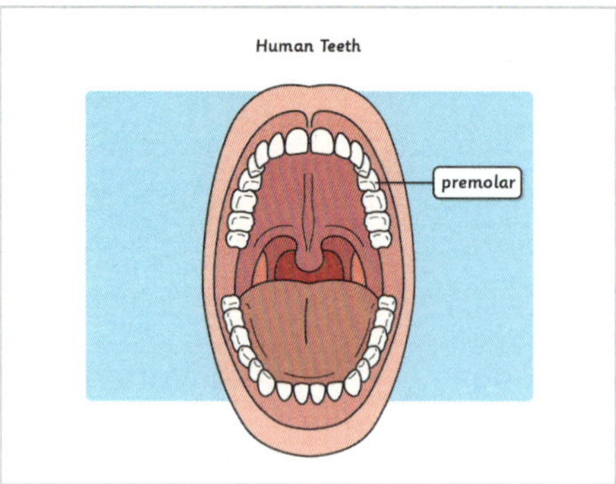

14. Tooth _ _ _ _ _ is when parts of a tooth are destroyed.

a b c

- decay
- dicay
- dekay
- dikay
- deca
- deka

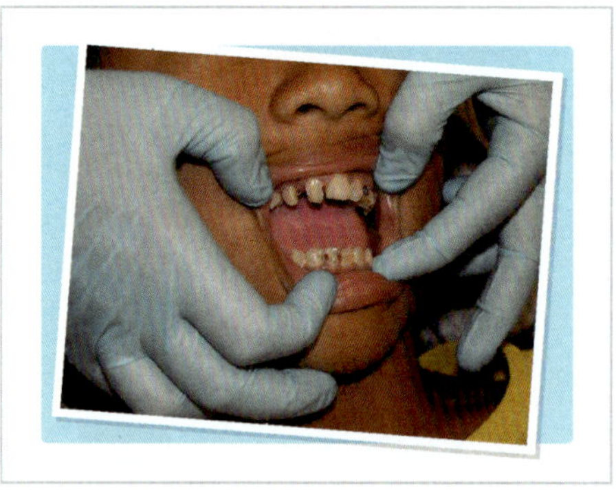

The Human Digestive System

Objective: I can describe the simple functions of the basic parts of the human digestive system.

Quick Search Ref: 10452

Level 1: Knowledge: I can identify parts of the human digestive system.

Required: 8/8 **Pupil Navigation:** on **Randomised:** off

1. What does our digestive system do?

1/3

- Pumps blood around the body.
- **Breaks down food and takes in the nutrients that our bodies need.**
- Produces offspring.

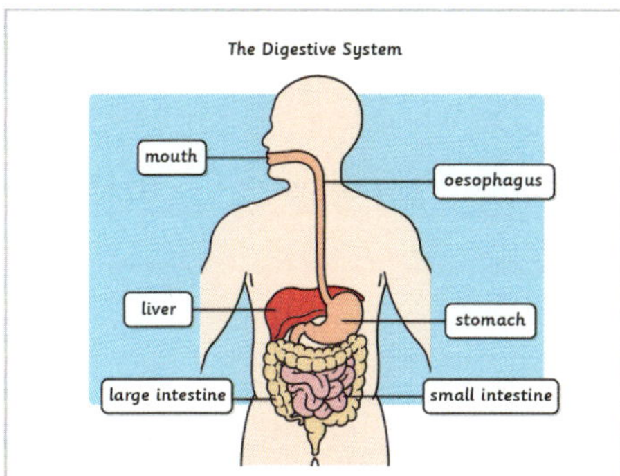

2. Select the **three** parts of the body that are involved in digestion.

3/5

- **stomach**
- **mouth**
- heart
- **teeth**
- lungs

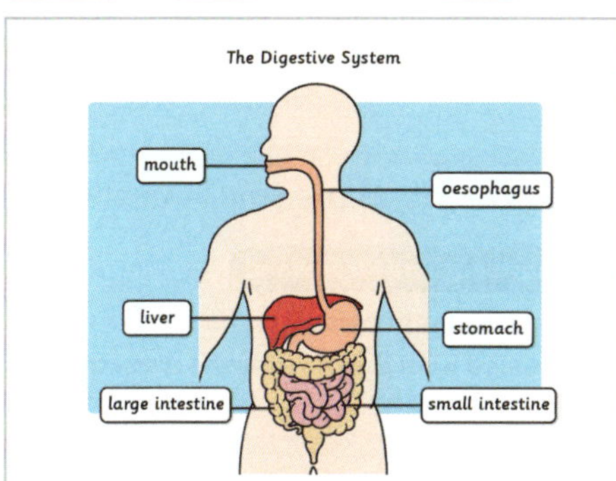

3. Which label points to the stomach?

1/5

- A
- B
- C
- **D**
- E

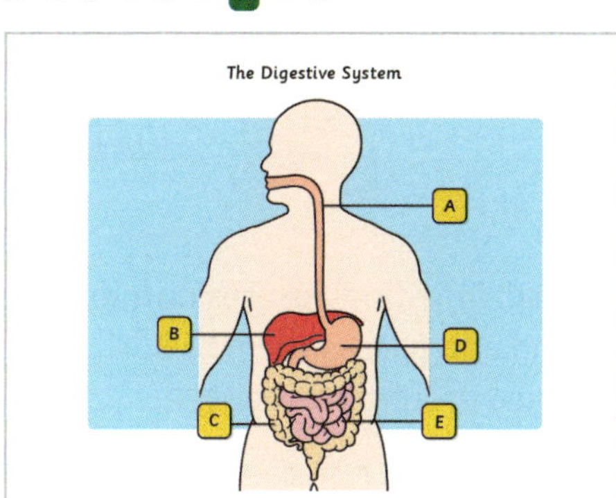

4. Which part of the digestive system does label A point to?

1/3

- windpipe
- small intestine
- **oesophagus**

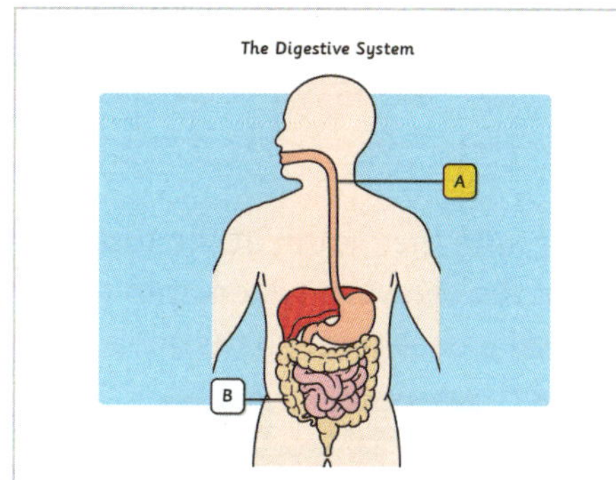

5. Which label points to the large intestine?

1/5

- A
- B
- **C**
- D
- E

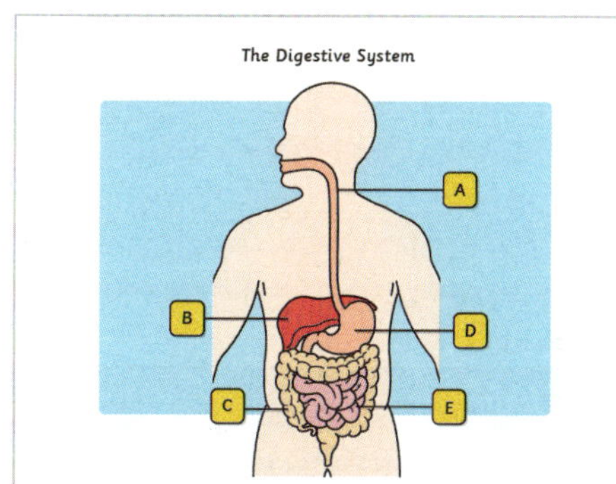

6. Which part of the digestive system does the label point to?

1/3

- large intestine
- anus
- **small intestine**

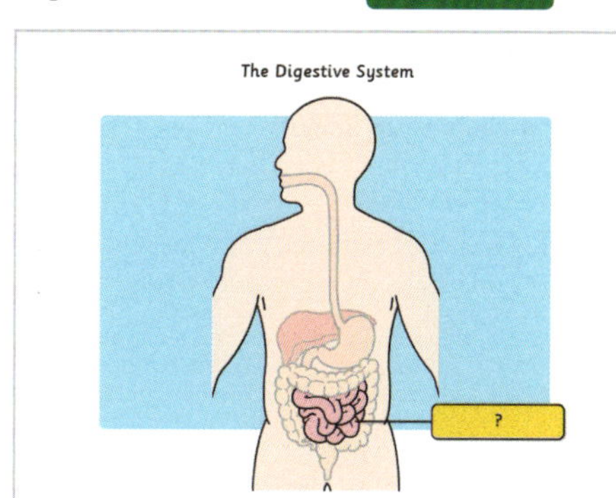

Level 1 *continued*

7. Which label points to the anus?

1/5

- A
- B
- C
- D
- **E**

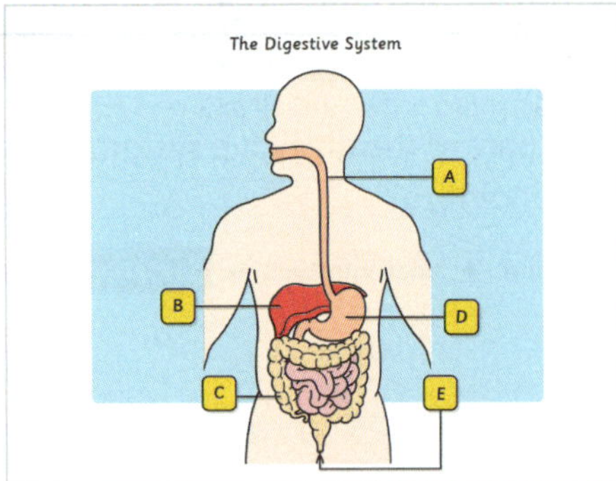

8. Starting with the mouth at the top, put the parts of the digestive system into the order that food passes through them.

- **mouth**
- **oesophagus**
- **stomach**
- **small intestine**
- **large intestine**
- **anus**

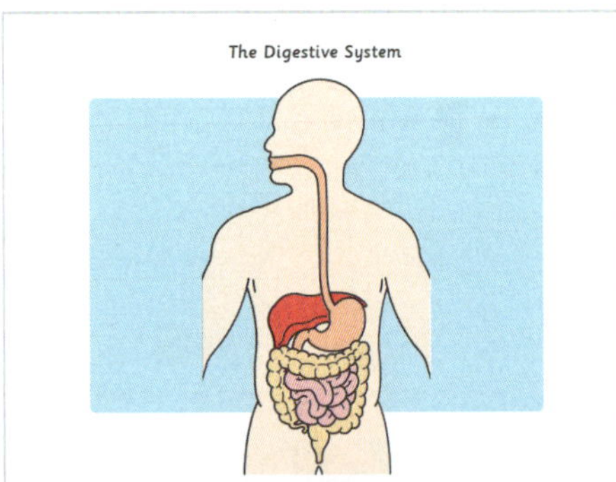

Level 2: Understanding: I can describe the functions of different parts of the digestive system.

Required: 8/8 **Pupil Navigation:** on

Randomised: off

9. Saliva starts to break down food in the _ _ _ _ _.

- mouf.
- mowth
- **mouth**
- **mouth.**
- mowf.
- mouf
- mowth.
- mowf

10. The muscles in the oesophagus . . .

1/3

- **push food from the mouth to the stomach.**
- help us to swallow food.
- use acid to break down food.

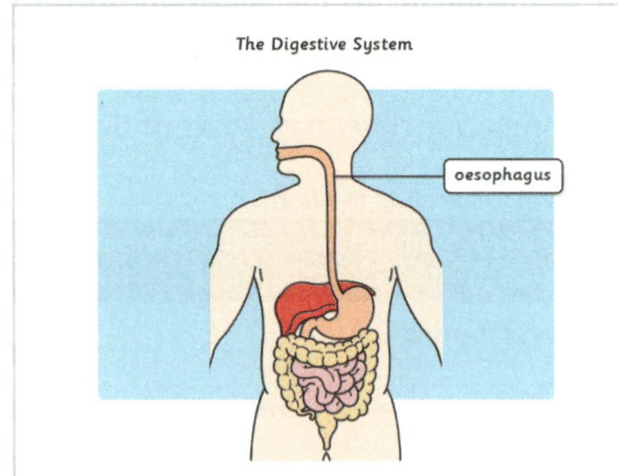

11. Your stomach contains _ _ _ _ to kill bacteria in food.

- **acid**
- asid

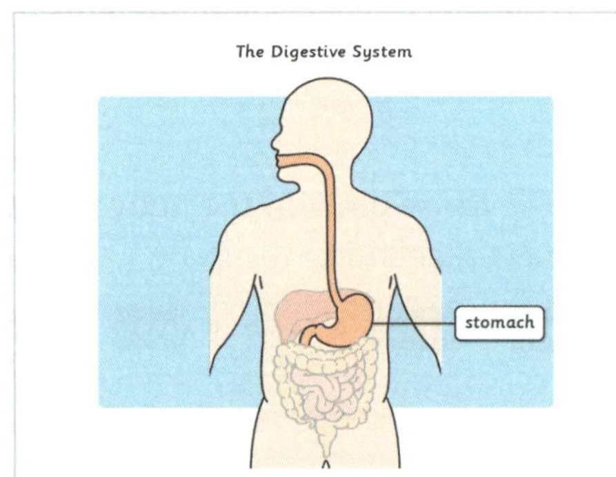

12. What is the job of the small intestine?

1/3

- **To absorb nutrients from food.**
- To absorb water from food.
- To pass food from the throat to the stomach.

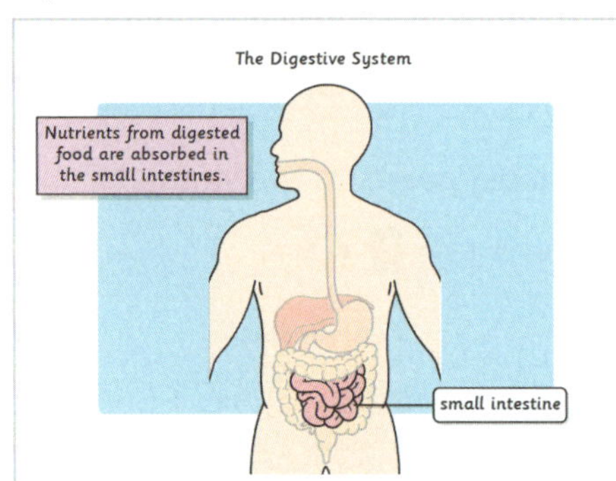

13. The small intestine absorbs nutrients from our food and passes them to our _ _ _ _ _ to be carried around the body.

- bloode
- **blood**
- blod
- dloob
- blude
- blud

Level 2 *continued*

14. Which part of the digestive system absorbs water?

1/4

- stomach
- small intestine
- **large intestine**
- anus

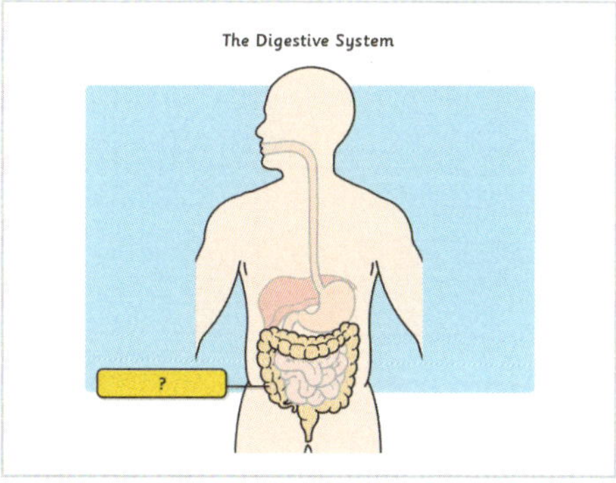

15. The anus is where...

1/3

- **undigested food leaves the body.**
- food is broken down.
- food enters the body.

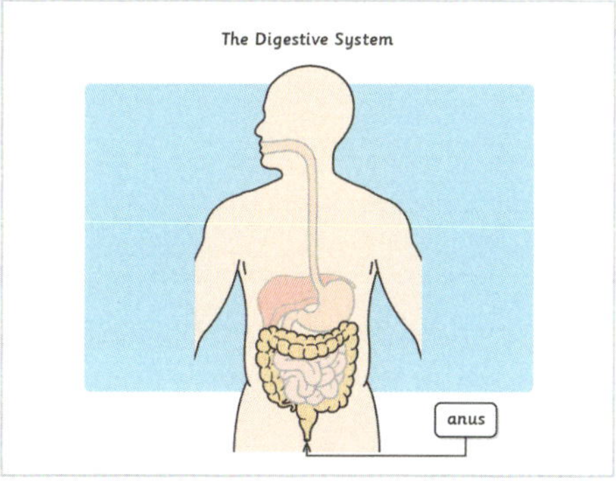

16. Faeces is the scientific word for . . .

1/3

- **the undigested food that leaves the body.**
- the acid that breaks down food in our stomachs.
- the bowel.

Level 3: Application: I can apply my knowledge and understanding of the digestive system to the world around me.

Required: 4/4 **Pupil Navigation:** on
Randomised: off

17. Some animals that eat grass, such as cows and rabbits, pass their food through their digestive systems more than once. Jenna, Mike and Stacey are discussing the reason why. Who is correct?

1/3

- **Mike**
- Jenna
- Stacey

18. Some animals partially digest food and then feed it to their young. This is called regurgitation. Why do you think they do this?

1/3

- To take all of the nutrients out of it.
- Because they're greedy.
- **Because their young can't fully digest their food.**

19. Jonny has a digestive disease and his body isn't absorbing enough nutrients from food. Which part of the digestive system do you think isn't working properly?

1/3

- **small intenstine**
- stomach
- large intestine

Level 3 *continued*

20. Your stomach is approximately the size of a . . .

☐ ☒ ☐ 1/3

- football.
- clenched fist.
- grape.

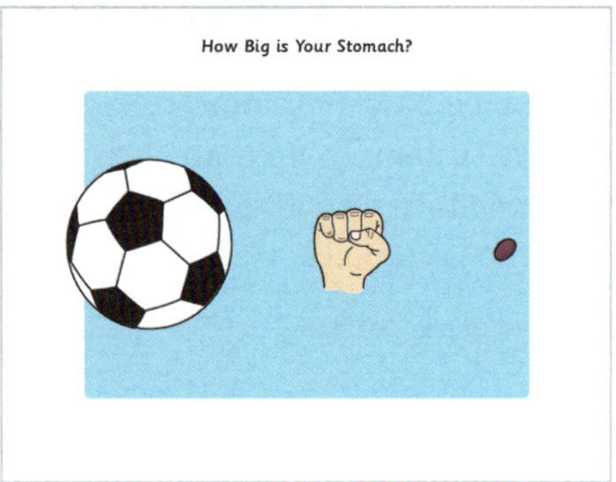

Types and Functions of Teeth

Objective: I can identify the different types of human teeth and know their simple functions.

Quick Search Ref: 10455

Level 1: Knowledge: I can identify and describe the different types of teeth in humans.

Required: 10/10 | **Pupil Navigation:** on | **Randomised:** off

1. What do we use our teeth for?

- tasting food
- swallowing food
- **ripping and chewing food**

1/3

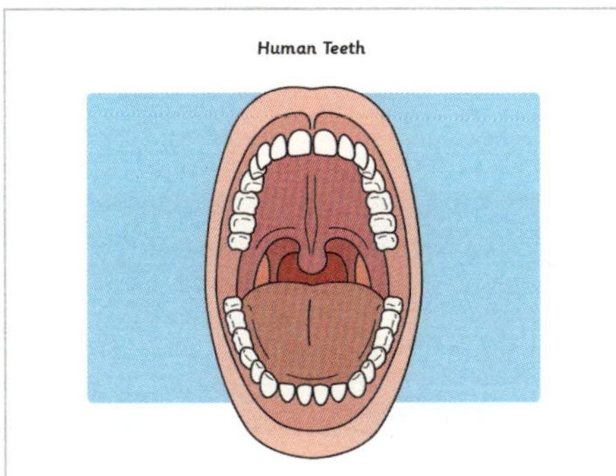

2. How many different types of teeth are there?

- **4**
- 3

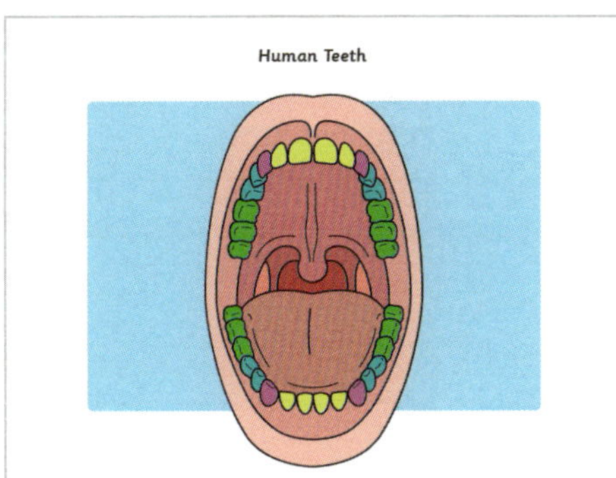

3. Which mouth shows a full set of adult teeth?

- A
- **B**

1/2

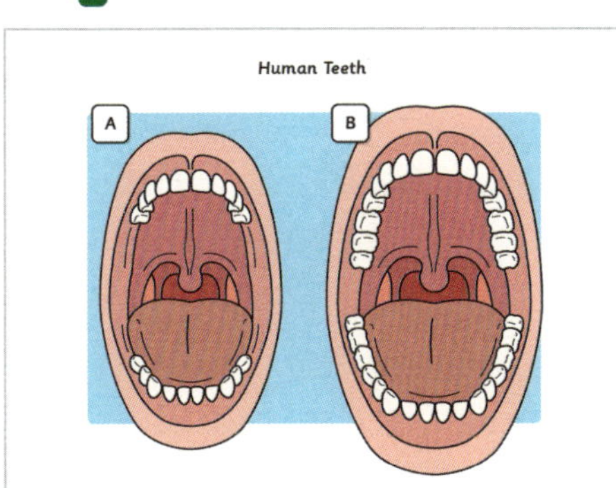

4. How many teeth are in a full set of milk teeth?

- **20**
- 32

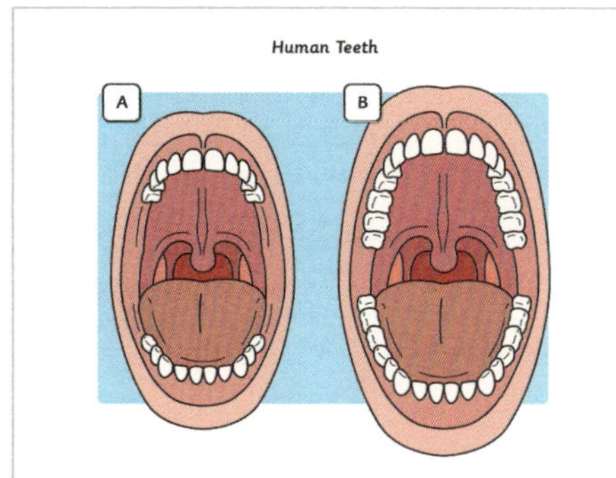

5. Which label points to a canine tooth?

- A
- **B**
- C
- D

1/4

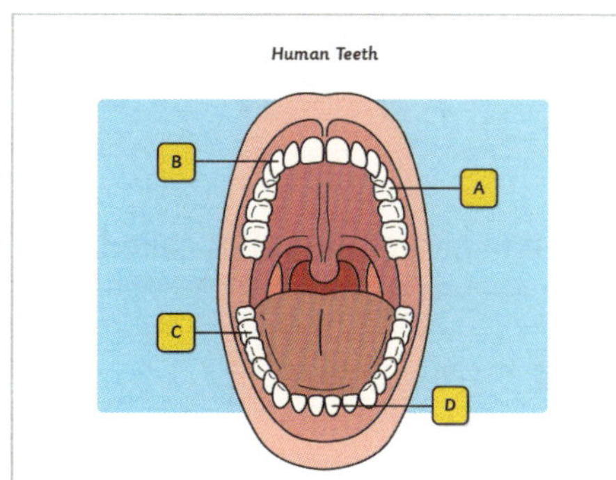

6. Which label belongs in the box?

- molars
- **premolars**
- canines
- incisors

1/4

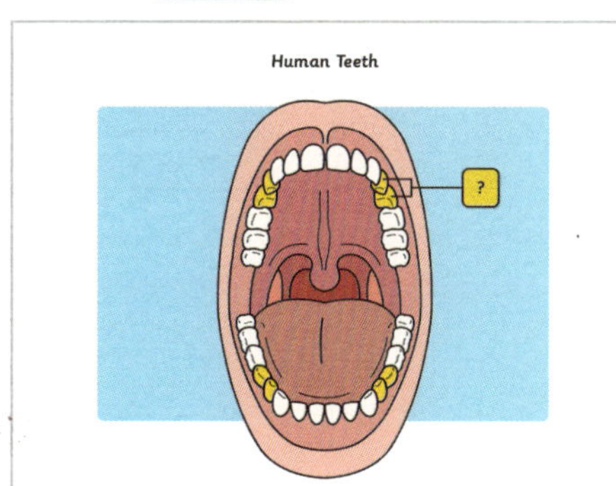

Level 1 *continued*

7. Label __ points to an incisor.

1/4

- W
- X
- **Y**
- Z

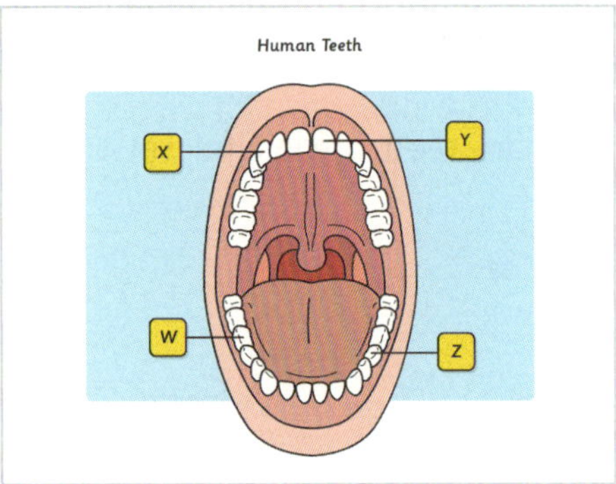

8. Label S points to which type of tooth?

1/4

- incisor
- canine
- premolar
- **molar**

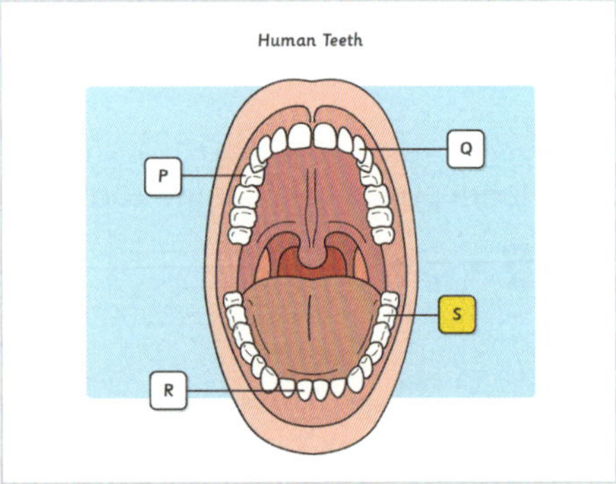

9. The visible part of a tooth is called the ______.

1/2

- **crown**
- root

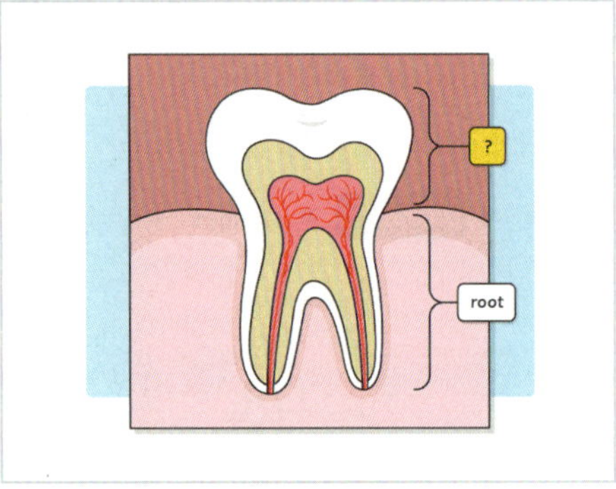

10. Sort the labels for the tooth from 1 to 5.
Label 1 goes at the top of the list.

- enamel
- dentine
- gum
- nerve
- pulp

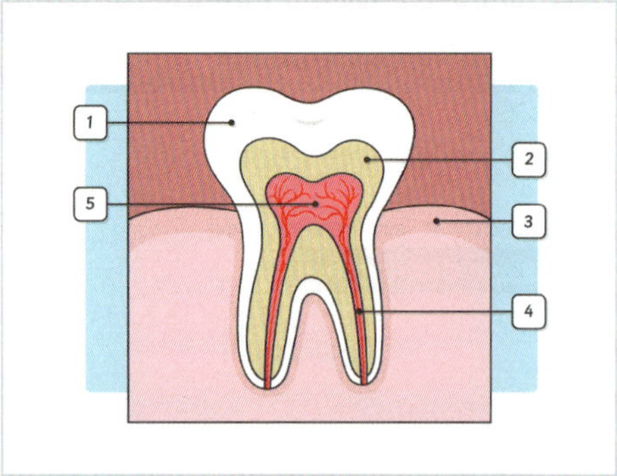

Level 2: Understanding: I can describe the functions of different human teeth.

Required: 7/7 **Pupil Navigation:** on

Randomised: off

11. What do we use our canine teeth for?

1/4

- To bite off pieces of food.
- To grind and chew food.
- **To grip, shred and tear foods such as meat.**
- To crush and tear food.

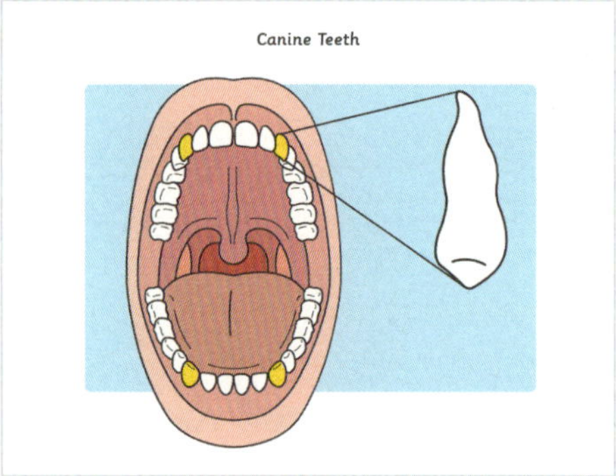

12. We use our ______ for grinding and chewing food.

- **molars**
- mowlars
- mowlas
- molas
- moles

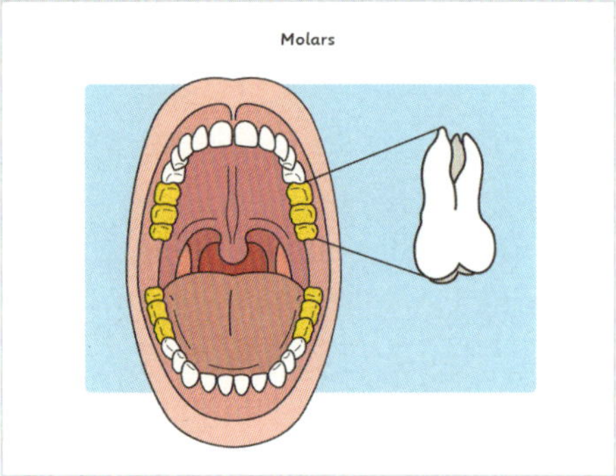

Level 2 *continued*

13. What do we use our incisors for?

1/4

- For gripping, shredding and tearing foods such as meat.
- **To bite off pieces of food.**
- For crushing and tearing food.
- For grinding and chewing food.

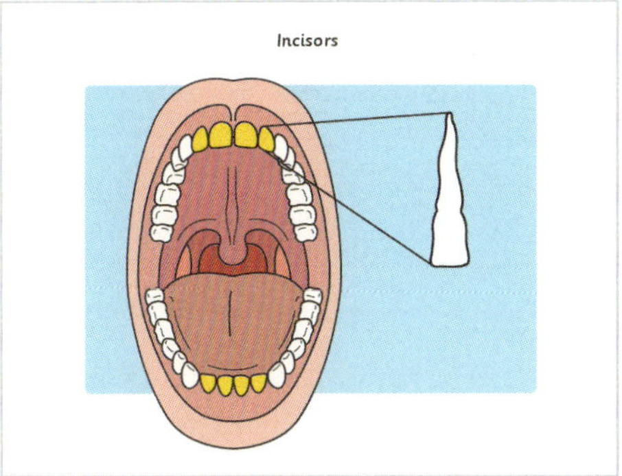

14. We use our premolars for . . .

1/4

- **crushing and tearing food.**
- gripping, shredding and tearing foods such as meat.
- biting off pieces of food.
- grinding and chewing food.

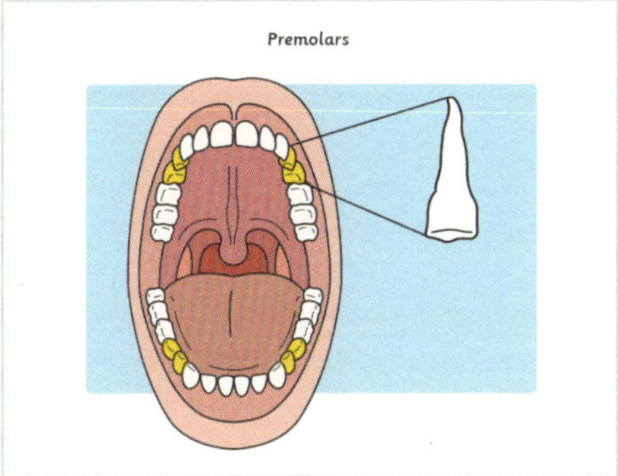

15. Casey, Araan and George are discussing why humans have four different types of teeth. Who do you agree with?

1/3

- Casey
- **Araan**
- George

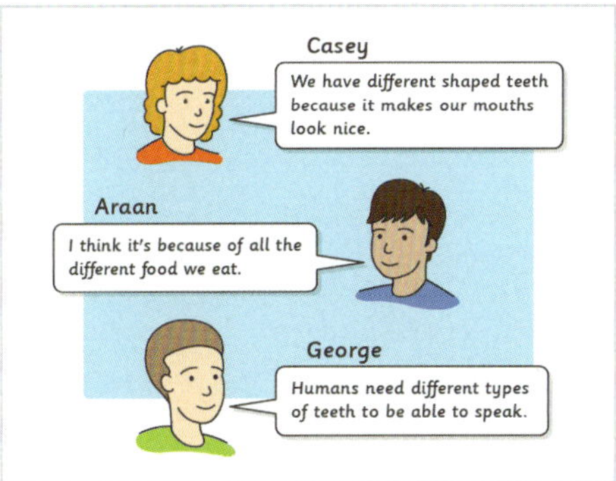

16. Which of these things help to keep our teeth healthy?

1/5

- going to the dentist
- brushing our teeth every day
- eating a healthy, balanced diet
- **all of these things**
- none of these things

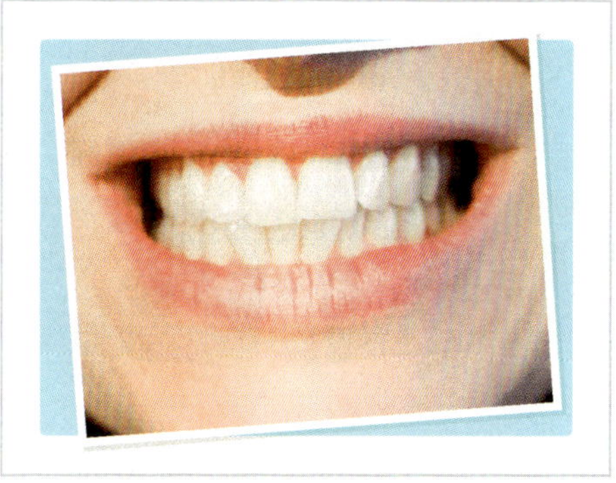

17. Your toothpaste contains ________ to help prevent tooth decay.

1/3

- mint
- water
- **fluoride**

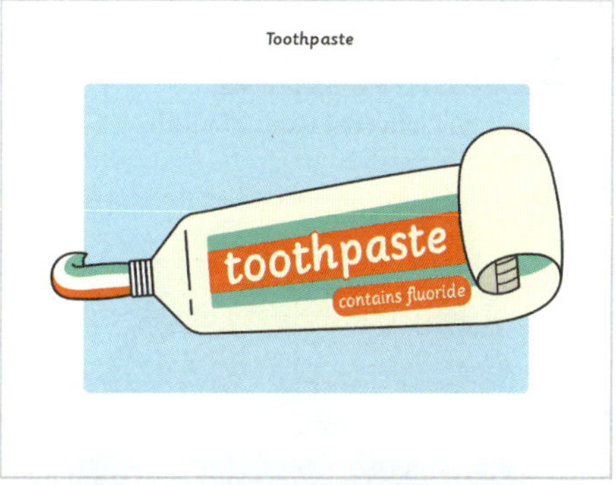

Level 3: Application: I can apply my knowledge and understanding of teeth and their functions to the world around me.

Required: 1/4 **Pupil Navigation:** on
Randomised: off

18. Suzy, Jonah and José are looking at a picture of an animal's jaw to try and work out what it eats. They can see **canines**, **incisors** and **molars**. Who do you agree with?

1/3

- Suzy
- Jonah
- **José**

Level 3 *continued*

19. Which jaws belong to a herbivore?

2/4

- A
- B
- C
- D

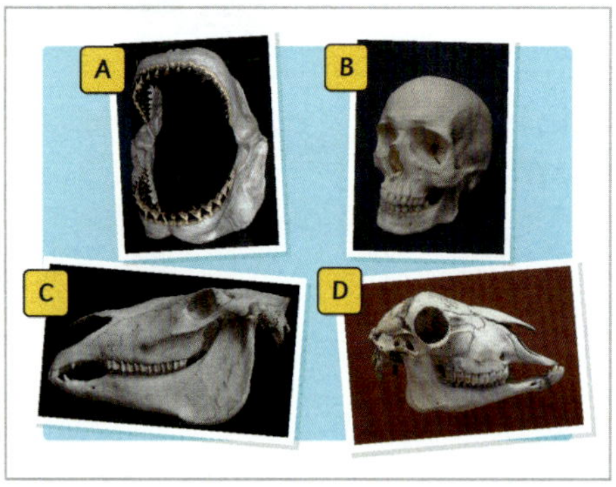

20. Palaeontologists can work out what dinosaurs are likely to have eaten by looking at fossils of their jaws. Why do they think that the tyrannosaurus rex was a carnivore?

1/4

- Because it ate other animals.
- Because it was bigger than all the other dinosaurs.
- Because it only had sharp teeth.
- Because it had different types of teeth.

21. Many animals, including humans, get a second set of teeth when they get older. Why do you think this is?

1/3

- Because when you are young, your mouth is too small for a full set of teeth.
- Because your diet changes as you get older, so you need stronger teeth.
- Because your first teeth fall out, so they need replacing.

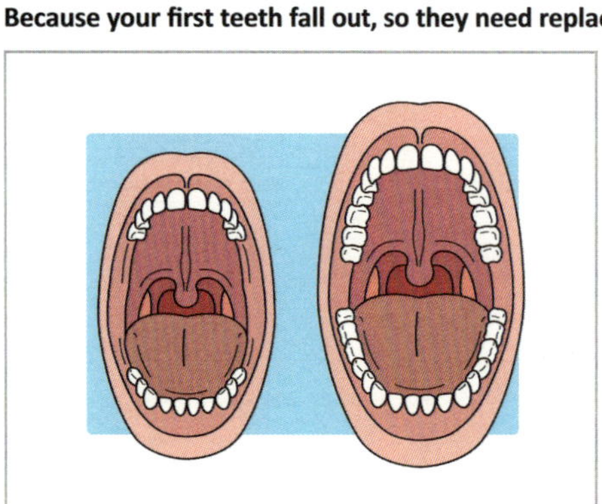

Primary Science Y3/4

Organisms

Nutrition

Food and Diet: Food Groups

Objective: I know what the food groups are, why we need them and which foods are good sources of them.

Quick Search Ref: 11498

Level 1: Knowledge: I can name the main food groups and know why they are needed.

Required: 10/13 **Pupil Navigation:** on **Randomised:** off

1. Why do we need nutrients from food?

1/2

- so that our food tastes nice
- so that our bodies can function properly

We get nutrients from the foods we eat.

2. Foods can be sorted into groups with other foods that contain a lot of the same _ _ _ _ _ _ _ _.

- nutrient
- nutrients

Foods in Different Food Groups

nutrient groups	food in this group
carbohydrates	potato, rice, bread, sweets
protein	meats, chicken, nuts, cheese
fats	butter, olive oil, oily fish
fibre	seeds, vegetables, cereals
vitamins and minerals	milk, cheese, fruit, vegetables

3. Which food group helps you grow bigger and repairs damage to your body?
Use the table to find the answer.

1/5

- carbohydrates
- protein
- fats
- fibre
- vitamins and minerals

Why is Each Food Group Needed?

food group	why it is needed
carbohydrates	energy
protein	growth and repair
fats	energy and to keep warm
fibre	keeps digestive system healthy
vitamins and minerals	keeps body healthy and working properly

4. Which food group helps you stay warm and gives you energy?

1/5

- carbohydrates
- protein
- fat
- fibre
- vitamins and minerals

Why is Each Food Group Needed?

food group	why it is needed
carbohydrates	energy
protein	growth and repair
fats	energy and to keep warm
fibre	keeps digestive system healthy
vitamins and minerals	keeps body healthy and working properly

5. Which two groups include foods that give you energy?

2/5

- carbohydrates
- protein
- fats
- fibre
- vitamins and minerals

Why is Each Food Group Needed?

food group	why it is needed
carbohydrates	energy
protein	growth and repair
fats	energy and to keep warm
fibre	keeps digestive system healthy
vitamins and minerals	keeps body healthy and working properly

Level 1 *continued*

6. Why does your body need fibre?

1/5

- for energy
- for energy and to keep warm
- for growth and repair
- **to keep the digestive system healthy**
- to keep the body healthy and working properly

Why is Each Food Group Needed?

food group	why it is needed
carbohydrates	energy
protein	growth and repair
fats	energy and to keep warm
fibre	keeps digestive system healthy
vitamins and minerals	keeps body healthy and working properly

7. Which food group helps you stay healthy and keeps everything in your body working properly?

1/5

- carbohydrates
- protein
- fats
- fibre
- **vitamins and minerals**

Why is Each Food Group Needed?

food group	why it is needed
carbohydrates	energy
protein	growth and repair
fats	energy and to keep warm
fibre	keeps digestive system healthy
vitamins and minerals	keeps body healthy and working properly

8. Calcium is a mineral that is important for strong _ _ _ _ _ and teeth.

- **bones**
- bowns

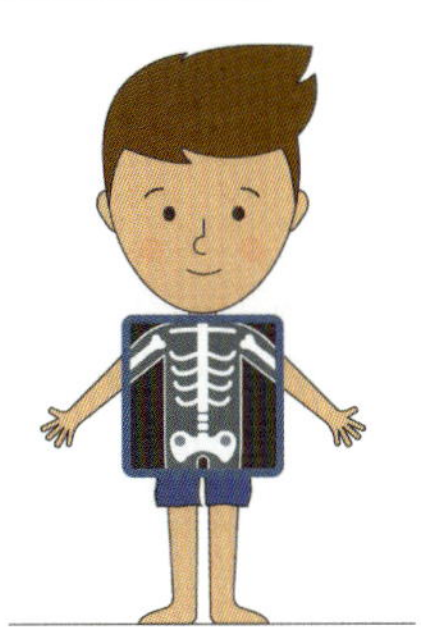

9. You need to eat the right foods and drink lots of _ _ _ _ _ to stay healthy.

- **water**
- warter
- watar

10. You need to eat lots of carbohydrates if you're doing lots of _________ .

1/3

- **exercise**
- singing
- sleeping

11. Your body needs protein so that it can _______ .

1/3

- **grow**
- get energy
- stay warm

Level 1 *continued*

12. Fats keep you _ _ _ _ and also give you energy.

- warm

Why is Each Food Group Needed?

food group	why it is needed
carbohydrates	energy
protein	growth and repair
fats	energy and to keep warm
fibre	keeps digestive system healthy
vitamins and minerals	keeps body healthy and working properly

13. Which system in your body needs fibre to stay healthy?

1/3

- digestive system
- nervous system
- circulatory system

Level 2: Understanding: I can give examples of foods that are good sources of different nutrients.

Required: 14/14 **Pupil Navigation:** on
Randomised: off

14. What does it mean if we say that a food is a **good source** of carbohydrates?

1/2

- The food only contains carbohydrates and no other nutrients.
- We can get a lot of carbohydrates from the food compared to some other foods.

15. Fruits and vegetables are good for your digestive system. What are they **good sources** of?

1/4

- carbohydrates
- fibre
- protein
- fats

16. Calcium is important for healthy, strong bones. You get calcium when you drink _ _ _ _, or when you eat cheese or yogurt.

- milk

17. Foods made from potato or flour are good sources of carbohydrates. Which two foods are good sources of carbohydrates?

2/4

- baked potato
- butter
- bread
- fish

Level 2 *continued*

18. Some good sources of carbohydrates contain a lot of sugar and taste very sweet. Which of the foods shown is a good source of carbohydrates?

1/4

- carrots
- doughnut
- cheese
- fish

19. Which of the foods shown is **not** a good source of protein?

1/4

- nuts
- fish
- potatoes
- chicken

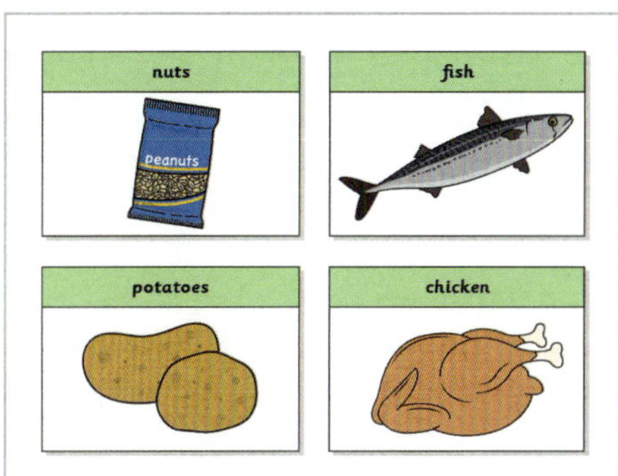

20. Vegetarians don't eat meat. Which three foods could a vegetarian eat to get protein in their diet?

3/5

- chicken
- nuts
- beef
- beans
- eggs

21. Some foods containing fats feel greasy or oily. Which **two** foods are good sources of fats?

2/4

- vegetable oil
- bread
- broccoli
- cheese

22. Fruits like oranges, lemons and limes are good sources of ________ .

1/3

- protein
- vitamins
- fats

23. On some breakfast cereal packets it says '*fortified with vitamins and minerals*'. What does this mean?

1/2

- Extra vitamins and mineral have been added.
- Vitamins and minerals have been removed.

Level 2 *continued*

24. Cheese and yogurt are made from milk. Which mineral are these foods good sources of?

1/3

- vitamin C
- **calcium**
- iron

25. Which of the foods shown is a good source of carbohydrates?

1/4

- fish
- **pasta**
- an orange
- cheese

26. Which nutrient are bread, pasta, rice and sweets good sources of?

1/5

- protein
- fats
- **carbohydrates**
- fibre
- vitamins and minerals

27. What are vegetables, seeds and cereals good sources of?

1/5

- protein
- fats
- carbohydrates
- **fibre**
- vitamins and minerals

Level 3: Application: I can analyse meals and design meals that contain sources of different food groups.

Required: 6/6 **Pupil Navigation:** on

Randomised: off

28. Marie has fish, chips and broccoli for her lunch. Which part of her meal is a good source of protein?

1/3

- **fish**
- chips
- broccoli

29. Which meal doesn't include a good source of carbohydrates?

1/3

- chicken curry with rice
- **cheese salad**
- spaghetti, tomato sauce and garlic bread

Level 3 *continued*

30. William is about to eat a burger on a bun when his mum suggests that he puts some tomato, lettuce and onion on it.
Which extra nutrients would William get by adding these three ingredients to his burger?

2/4

- carbohydrates
- **fibre**
- protein
- **vitamins**

31. Tanya is making grilled chicken for her dinner. What could she have with the chicken so that she also has a good source of carbohydrates?

1/4

- **rice**
- tomato
- peas
- cheese

32. What nutrients will Adam get if he eats a bowl of pasta with butter on it for his lunch?

- Open question, no set answer

33. Tom is making a cheese and tomato pizza for his lunch. The pizza base is a good source of carbohydrates, and the cheese is a good source of fats. What else could Tom add to his meal so that he gets more protein and fibre?

- Open question, no set answer

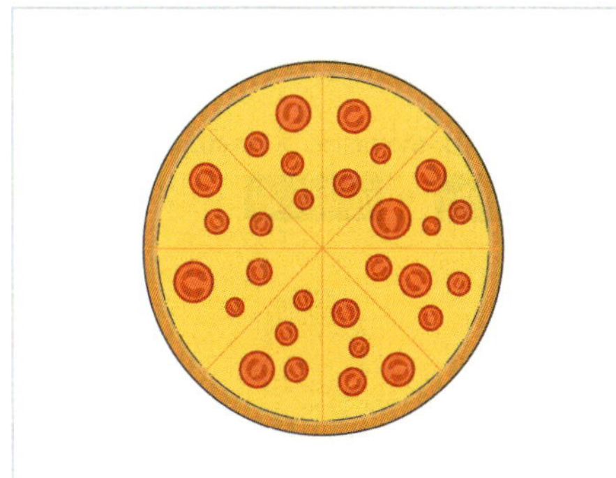

Food and Diet: Having a Balanced Diet

Objective: I know why a balanced diet is important, how to make my diet balanced and what can happen if it's not.

Quick Search Ref: 11486

Level 1: Knowledge: I know that a balanced diet is important for health and is different for different people.

Required: 8/8 **Pupil Navigation:** on **Randomised:** off

1. Why do our bodies need food?
Select three answers.

3/5

- to grow
- for the taste
- to stop us feeling hungry
- to stay healthy
- for energy

2. We get carbohydrates, protein, fat, fibre, vitamins and minerals from the foods we eat. What is the correct name for these things?

1/3

- substances
- nutrients
- ingredients

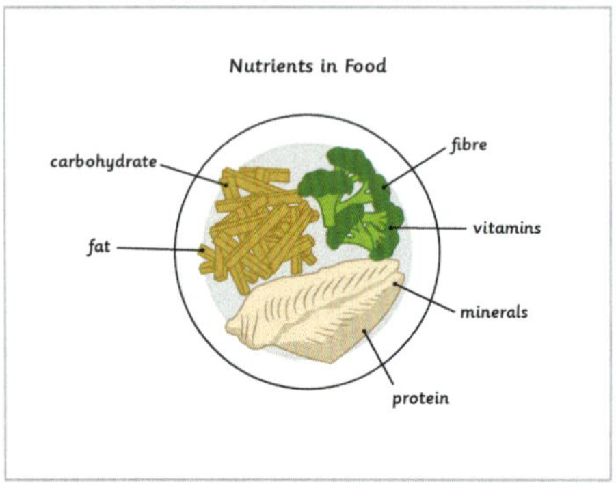

3. Does every food contain all of the nutrients we need for energy, growth and good health?

1/2

- Yes
- No

4. What does **diet** mean?

1/2

- Diet is the word for what you eat when you want to lose weight.
- Diet is the word for all of the foods that you regularly eat.

5. What do you need to eat to have a healthy balanced diet?

1/3

- similar amounts of different foods
- the same amount of each type of nutrient
- a mixture of foods so that your body gets the right nutrients

Level 1 *continued*

6. How many portions of fruit and vegetables should you try to eat each day to stay healthy?

1/4

- one portion a day
- five portions a day
- ten portions a day
- three portions a day

7. True or false? As long as you eat foods that give you the right nutrients, it doesn't matter how much of them you eat.

1/2

- True
- False

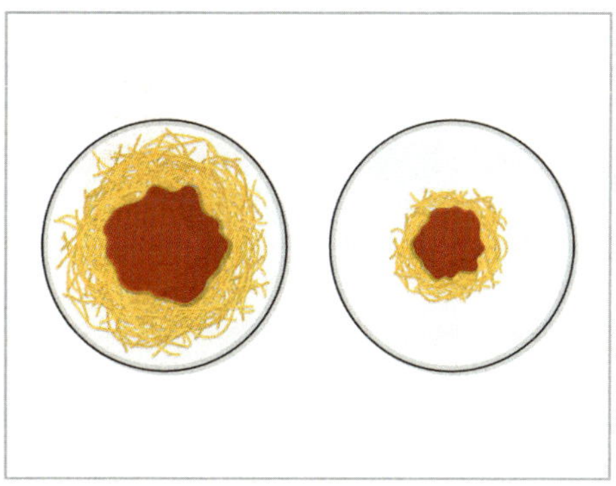

8. Select **two reasons** why a healthy balanced diet for you isn't the same as a healthy balanced diet for your teacher.

2/4

- Your teacher has stopped growing, but you still need nutrients to grow.
- You and your teacher do different amounts of exercise, so you need different amounts of energy from your diets.
- You and your teacher like different types of foods.
- You have to eat what you're given, but your teacher gets to choose what they want to eat because they're an adult.

Level 2: Understanding: I know how food labels, portion size and cooking methods can influence a balanced diet.

Required: 9/9 **Pupil Navigation:** on
Randomised: off

9. What does the nutrition label on a food packet tell you?

1/4

- how much of each nutrient is in the food
- how to prepare the food
- ingredients in the food
- how to store the food

Nutrition Facts

Serving size 1/2 cup (115g)
Serving per container about 4

Amount per serving	
Calories 250	**Calories from Fat** 130
	% Daily Value
Total Fat 14g	22%
Saturated 9g	45%
Cholestrol 55mg	18%
Sodium 75mg	3%
Total Carbohydrate 26g	9%
Fibre 0g	0%
Sugars 26g	
Protein 4g	

10. You can tell how healthy a food is by looking at the *traffic light* information on the nutrition label. If a food is healthy, what colour will you see a lot of?

1/3

- red
- amber
- green

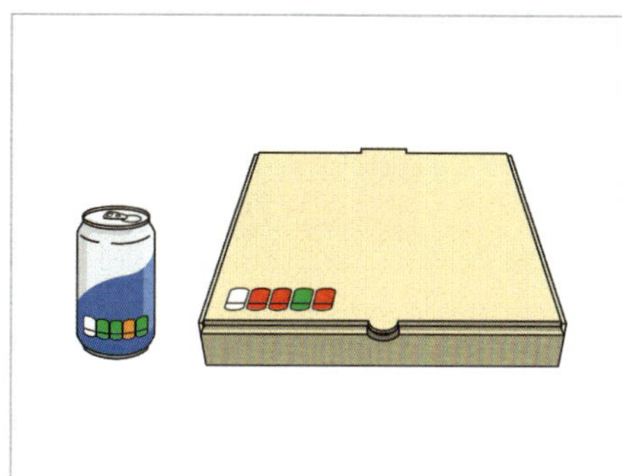

Level 2 *continued*

11. What two things does it tell you about a food if most of the nutrition label is red?

2/4

- It is a healthy food.
- **It is a food you should eat less often.**
- You should eat a large amount of the food.
- **You should only eat a small amount of the food.**

12. If the labels shown were on different packets of food, which food would be the healthiest choice?

1/4

- Food A
- Food B
- **Food C**
- Food D

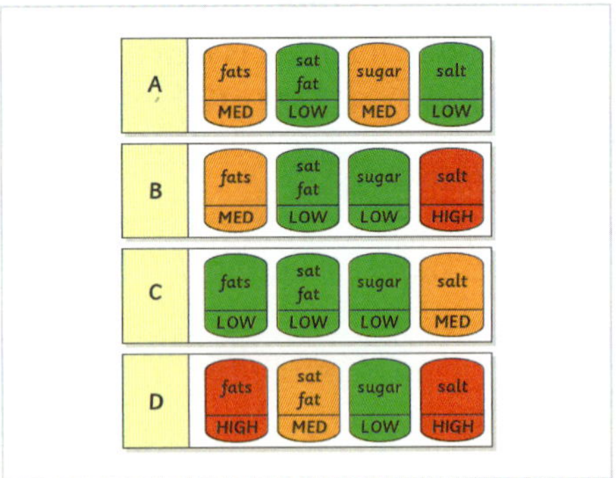

13. If the labels shown were on different packets of food, which food would be the least healthy choice?

1/4

- Food A
- Food B
- Food C
- **Food D**

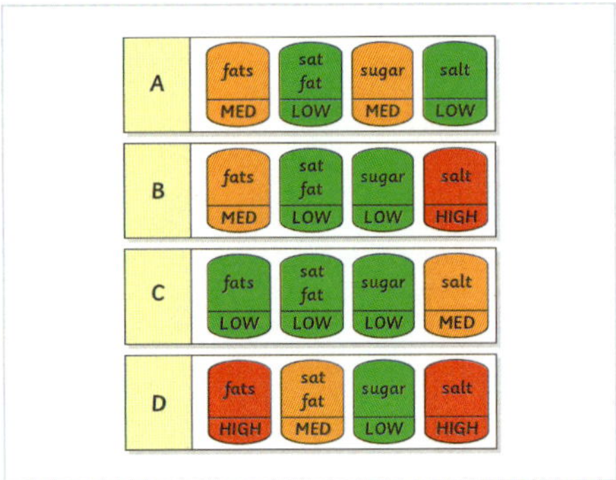

14. To get the nutrients described on the nutrition label, how much of this food would an adult need to eat?

1/3

- **half a pack**
- quarter of a pack
- a whole pack

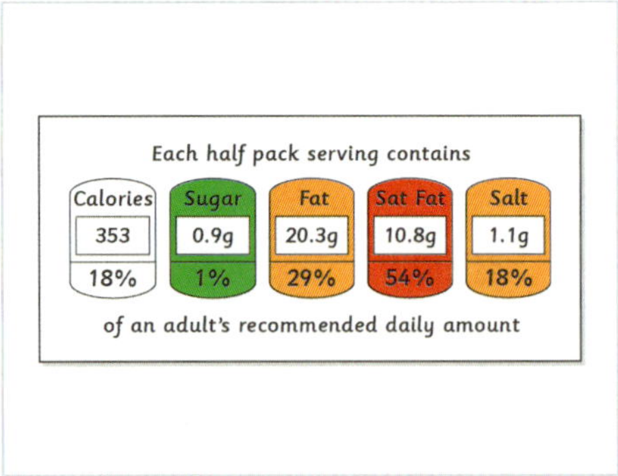

15. Does this label tell **a child** everything they need to know about the nutrients in this food? Explain your answer.

- Open question, no set answer

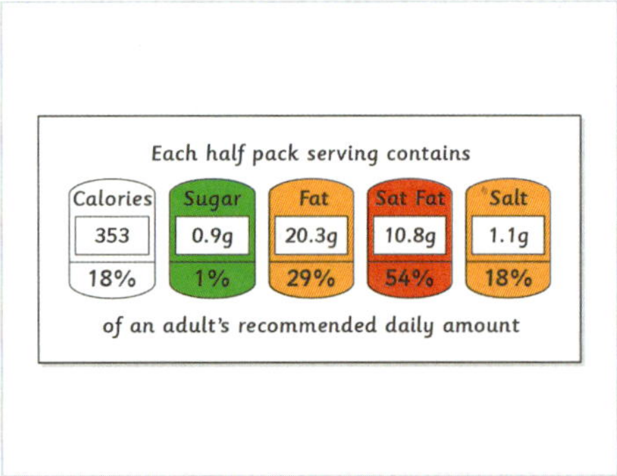

16. Why can it be hard to know if a takeaway meal is healthy or not?

- Open question, no set answer

Level 2 *continued*

17. True or false? The way food is cooked changes the nutrients you get from it.

1/2

- **True** ✓
- **False**

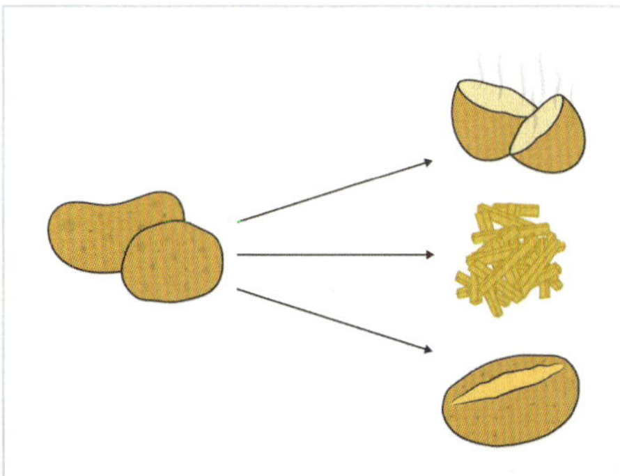

Level 3: Application: I can recognise and explain issues related to not having a balanced diet.

Required: 7/7 **Pupil Navigation:** on
Randomised: off

18. Abbie and Bella are identical twins. Identical twins usually grow to the same height, but Bella has grown taller than Abbie. What might this tell you about the amount of protein in Bella's diet?

1/3

- **Bella's diet had less protein in it.**
- **Bella and Abbie had the same amount of protein.**
- **Bella's diet had more protein in it.** ✓

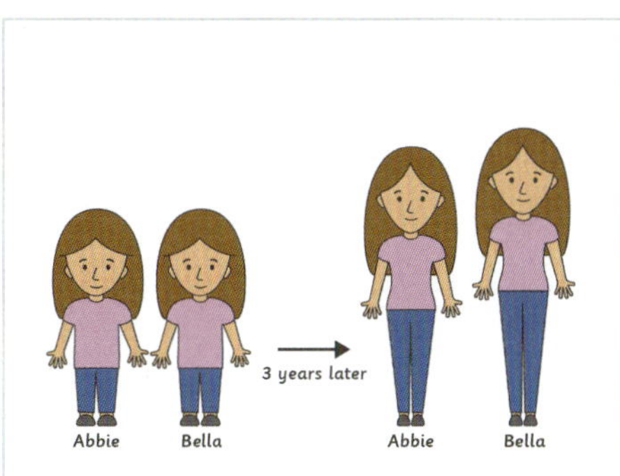

19. Alex wants to start exercising more but he doesn't feel like he has a lot of energy. How could he change his diet to give him more energy?

1/3

- **Alex could eat more carbohydrates for more energy.** ✓
- **Alex could eat more protein for more energy.**
- **Alex could eat more fibre for more energy**

20. Which disease can you get if you don't have enough vitamin C?

a b c

- **scurvy** ✓

If you don't get enough vitamin C in your diet, you could get a disease called scurvy.

Not many people get scurvy nowadays because we can get fresh fruit quite easily.

In the past, sailors often got scurvy because they were at sea for a long time without any fresh fruit in their diet.

21. Why did sailors get scurvy in the past?

1/3

- **They didn't eat enough fresh fruit.** ✓
- **They ate too much protein.**
- **They didn't drink enough fresh water.**

If you don't get enough vitamin C in your diet, you could get a disease called scurvy.

Not many people get scurvy nowadays because we can get fresh fruit quite easily.

In the past, sailors often got scurvy because they were at sea for a long time without any fresh fruit in their diet.

Level 3 *continued*

22. Which two foods could you eat to avoid getting scurvy?

2/4

- oranges
- cheese
- lemons
- chicken

If you don't get enough vitamin C in your diet, you could get a disease called scurvy.

Not many people get scurvy nowadays because we can get fresh fruit quite easily.

In the past, sailors often got scurvy because they were at sea for a long time without any fresh fruit in their diet.

23. Which two problems could you face if you ate too much fat?

2/4

- weak bones
- becoming overweight
- problems with your heart
- tooth decay

24. What problem can you face if you don't get enough calcium in your diet?

1/4

- chickenpox
- unhealthy skin
- weak bones
- strong teeth

Primary Science Y3/4

Ecosystems

Food Chains
Grouping and Classification
Habitats

Food Chain Vocabulary

Objective: I can use vocabulary relating to food chains.

Quick Search Ref: 10499

Level 1: Knowledge: I can spell and define scientific words for food chains.

Required: 10/10 **Pupil Navigation:** off **Randomised:** off

1. A _ _ _ _ _ _ _ _ _ _ is an animal that eats mainly meat.

abc

- carnibore
- **carnivore** (correct)
- herbivore
- carnevor
- omnivore
- carnevore
- carnivor

Carnivores

tiger, dolphin & vulture by Phillip Hartwell

2. A _ _ _ _ _ _ _ _ _ _ is an animal that eats mainly plants.

abc

- carnivore
- herbevore
- **herbivore** (correct)
- herbevor
- herbibore
- herbivor
- erbivore
- omnivore

Herbivores

giraffe & parrot by Phillip Hartwell

3. Omnivores are animals that eat . . .

1/4

- mainly other animals.
- mainly plants.
- **plants and other animals.** (correct)
- only fish.

Omnivores

badger by USFWS Mountain-Prairie; *wasp* by Trounce; *flamingo* by Phillip Hartwell

4. A _ _ _ _ _ _ _ _ kills and eats other animals.

abc

- **predator** (correct)
- predater
- preditor
- prediter
- prey

Animals That Hunt and Kill Their Prey for Food

tiger, dolphin & lion by Phillip Hartwell

5. Animals that are hunted and killed by other animals are known as _ _ _ _ animals.

abc

- **prey** (correct)
- pra
- predator
- prae
- pray
- herbivore

Animals That Are Hunted by Predators for Food

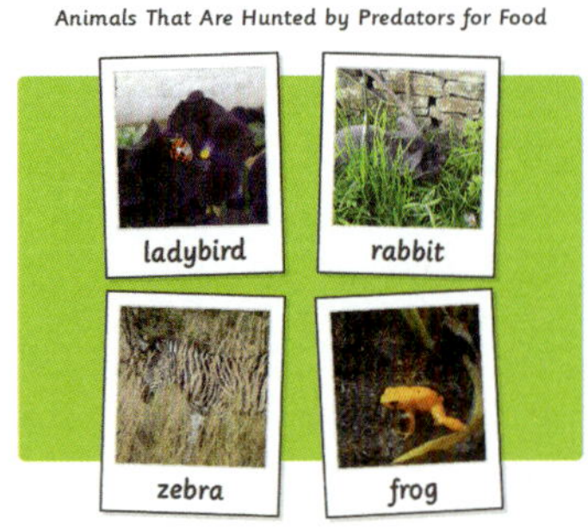

zebra by Phillip Hartwell

6. Vultures are birds that eat animals which are already dead. What is the name for animals who find their food in this way?

1/4

- predators
- carnivores
- prey
- **scavengers** (correct)

vulture by Phillip Hartwell

Level 1 *continued*

7. In a food chain, the _ _ _ _ _ _ _ _ _ makes its own food through photosynthesis.

- plant
- sun
- produsa
- producers
- producer
- plants
- prducer
- prey
- predator
- produser
- consumer

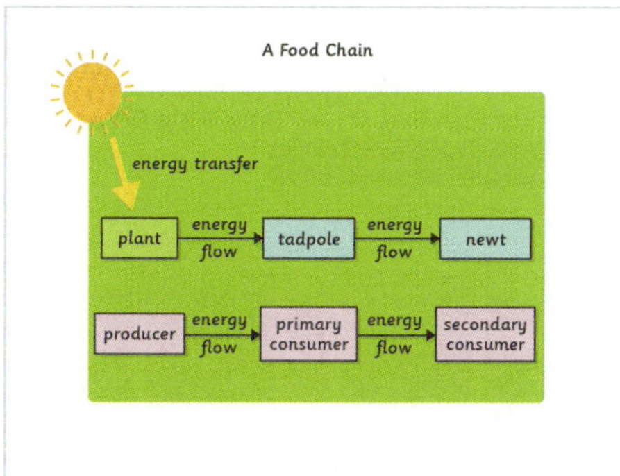

8. Animals can't make their own food, so they consume plants and other animals. In a food chain, animals are known as . . .

1/4

- producers.
- predators.
- consumers.
- prey.

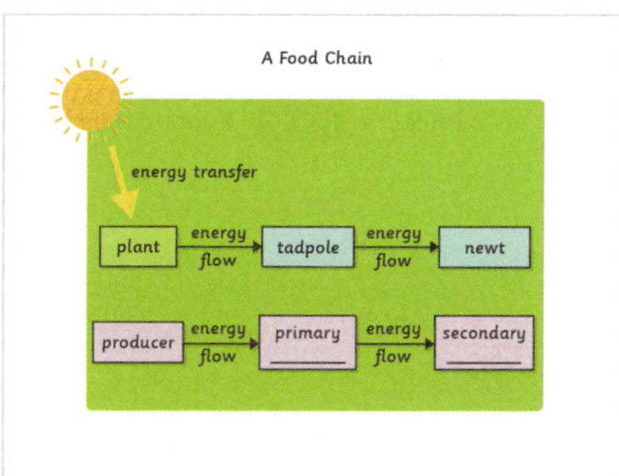

9. _ _ _ _ _ _ _ _ _ _ is the life process of taking in food needed for movement and growth.

1/3

- reproduction
- sensitivity
- nutrition

10. What does a food chain show?

1/3

- Which animals live in a habitat.
- The direction of energy flow between organisms in a habitat.
- The diets of different animals in a habitat.

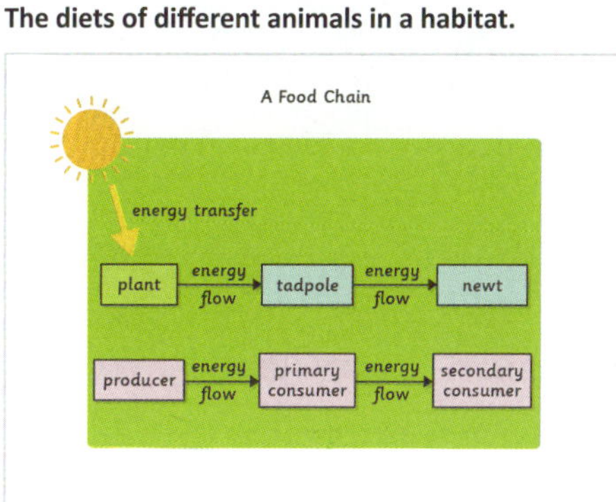

Parts of a Food Chain

Objective: I can construct and interpret a variety of food chains, identifying producers, predators and prey.

Quick Search Ref: 10548

Level 1: Knowledge: I can use food chains to identify producers, predators and prey.

Required: 9/12 | **Pupil Navigation:** on | **Randomised:** off

1. What is the plant in a food chain called?

1/4

- predator
- prey
- **producer**
- consumer

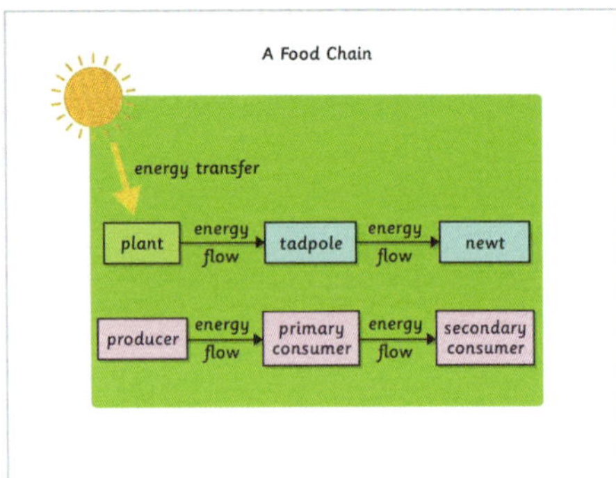

2. A predator is . . .

1/4

- an animal that eats animals that have been killed by other animals.
- a plant that can make its own food in its leaves.
- **an animal that kills and eats other animals.**
- an animal that is killed and eaten by other animals.

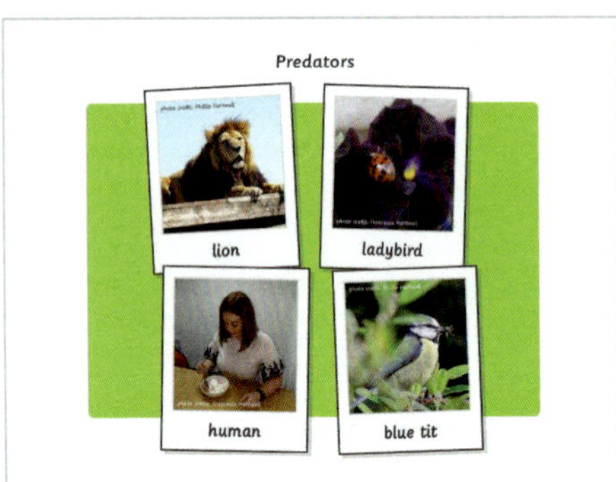

3. _ _ _ _ is the name for animals that are killed and eaten by other animals.

abc

- **prey**
- prae
- pray

4. Which statement is incorrect?

1/3

- Humans are predators.
- Prey is always an animal.
- **Predators are all carnivores.**

5. Which organisms are the predators in this food chain?

2/4

- **newt**
- tadpole
- **heron**
- pond weed

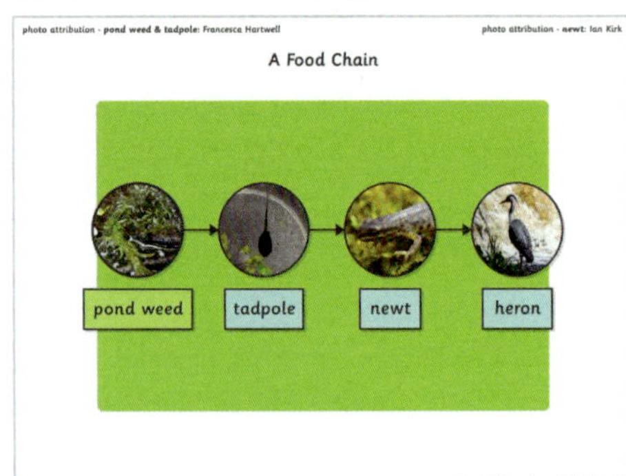

6. Which animals in these food chains are both prey and predators?

2/6

- cat
- **blackbird**
- tadpole
- heron
- caterpillar
- **newt**

Level 1 *continued*

7. A food chain shows how _ _ _ _ _ _ is transferred between living organisms.

■ energe ■ **energy** ■ energi ■ energie

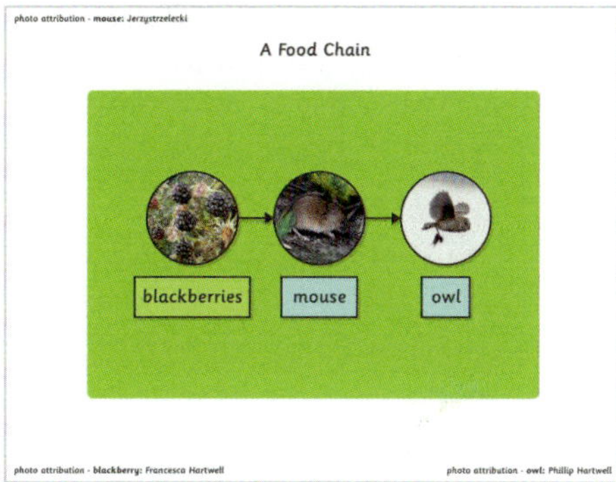

8. What do the arrows between the organisms in a food chain mean?

1/4

■ kills ■ eats ■ is killed by ■ **direction of energy flow**

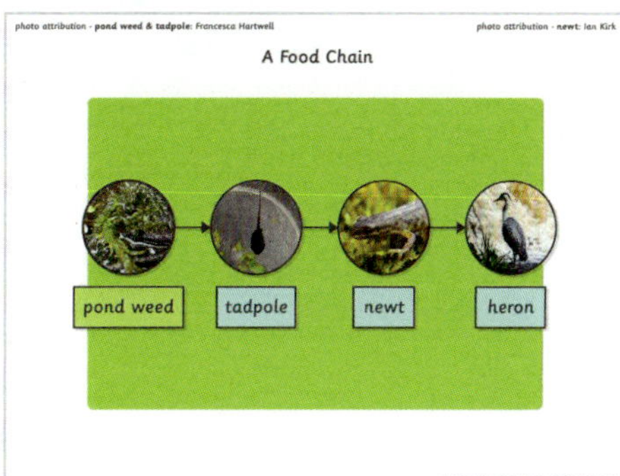

9. What is the name for an animal that isn't eaten by anything else in a food chain?

1/4

■ king predator ■ **top predator** ■ important predator ■ emperor predator

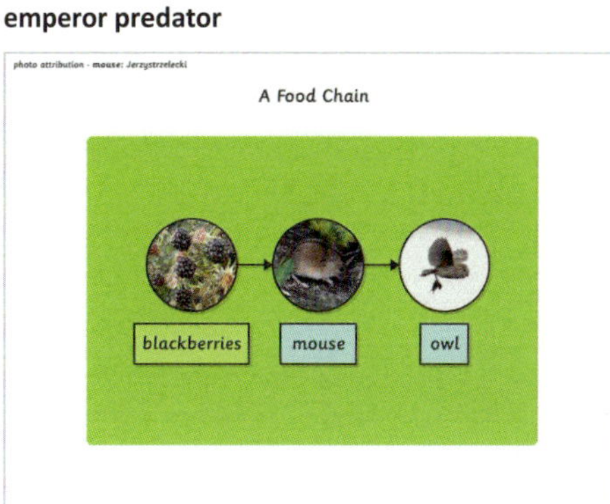

10. What is the producer in this food chain?

1/4

■ the Sun ■ blue tit ■ **ivy** ■ caterpillar

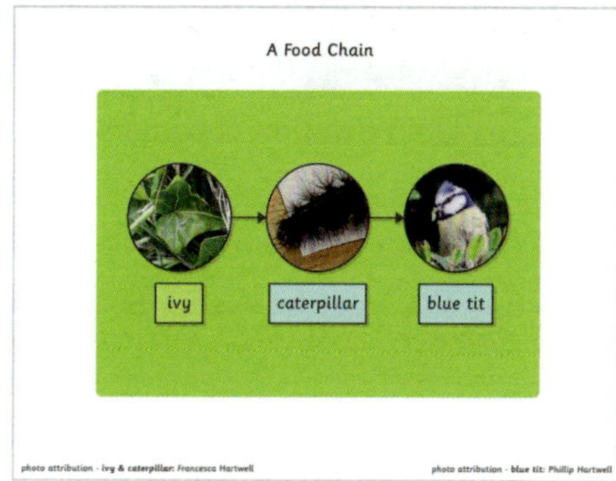

11. Which food chain is incorrect?

1/3

■ 1 ■ 2 ■ **3**

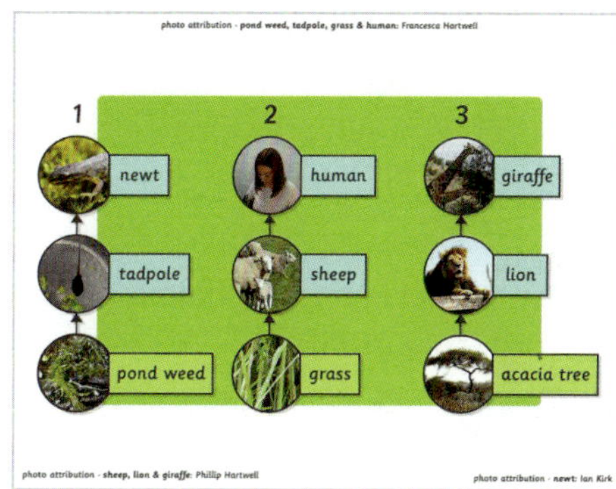

12. What is the prey in this food chain?

1/4

■ owl ■ the Sun ■ blackberries ■ **mouse**

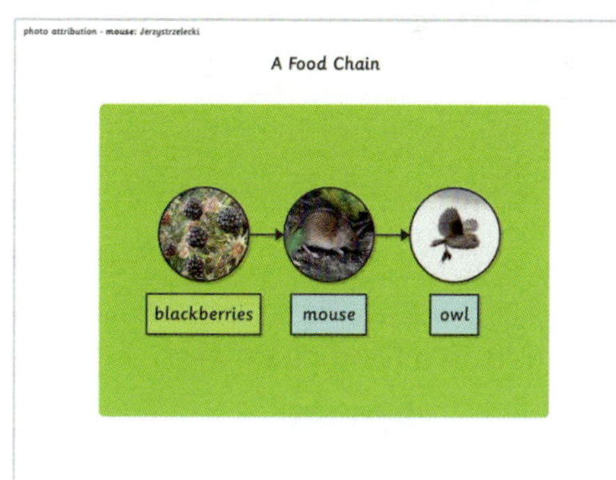

Level 2: Understanding: I can create food chains to identify producers, prey and predators.

Required: 6/7 **Pupil Navigation:** on
Randomised: off

13. Plants are called producers because . . .

1/3

- they are only eaten by herbivores.
- they are eaten by animals in the food chain that can't make their own food.
- **they produce their own food using air, water and light from the Sun.**

14. The first animal in a food chain is always a herbivore. True or false?

1/2

- true
- **false**

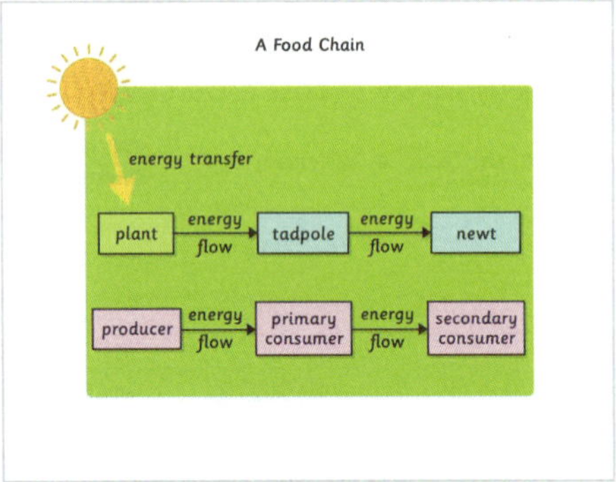

15. All animals in a food chain are ___________.

1/4

- scavengers
- **consumers**
- carnivores
- producers

16. Why is the Sun shown at the start of a food chain?

1/3

- **The plant uses light from the Sun to make its food.**
- We can always see the Sun.
- The Sun is a producer.

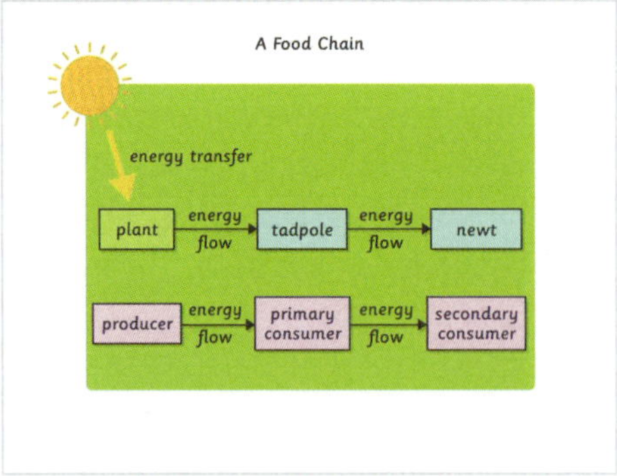

17. Which arrow should you use to complete the food chain?

1/4

- a
- b
- **c**
- d

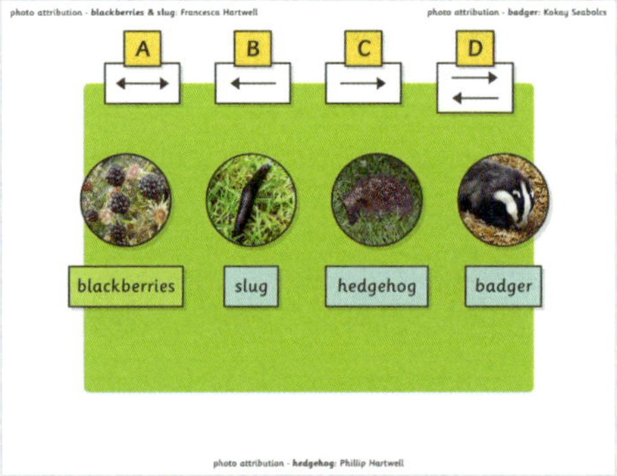

18. Starting with the producer, use the fact files to help you arrange the parts of this rainforest food chain into the correct order.

- **Sun**
- **brazil nut tree**
- **agouti**
- **jaguar**

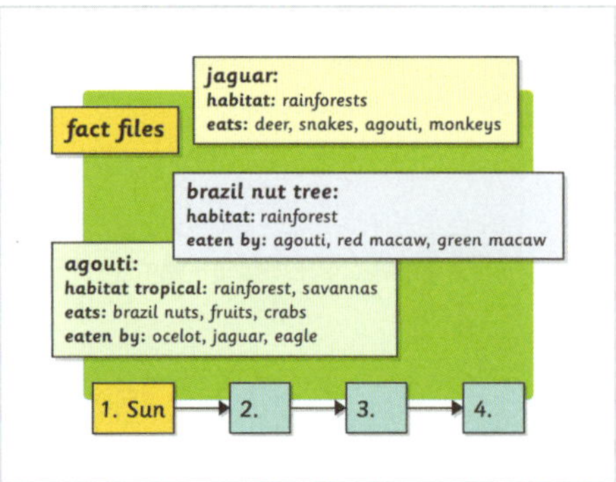

Level 2 *continued*

19. The Arctic is a very cold region around the North Pole. Which of these living things wouldn't be in an Arctic food chain?

2/5

- polar bear
- **penguin**
- plankton
- cod
- **palm tree**

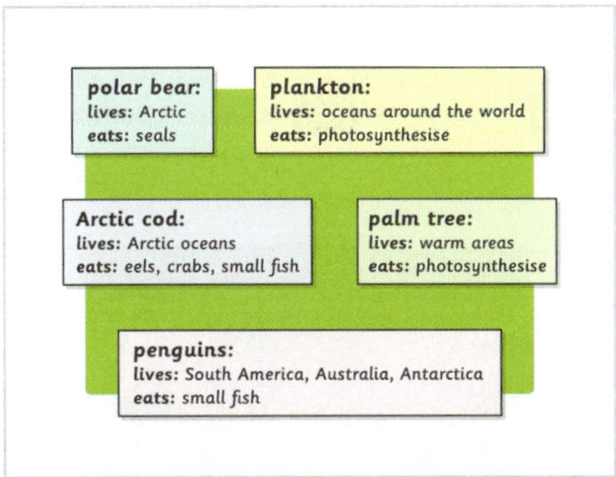

Level 3: Application: I can apply my knowledge and understanding of food chains to the world around me.

Required: 3/3 **Pupil Navigation:** on
Randomised: off

20. A food web shows how different food chains in a habitat link together. Select **two** animals that are prey in this woodland food web.

2/7

- falcon
- fox
- **caterpillar**
- **rabbit**
- badger
- **mouse**
- **sparrow**

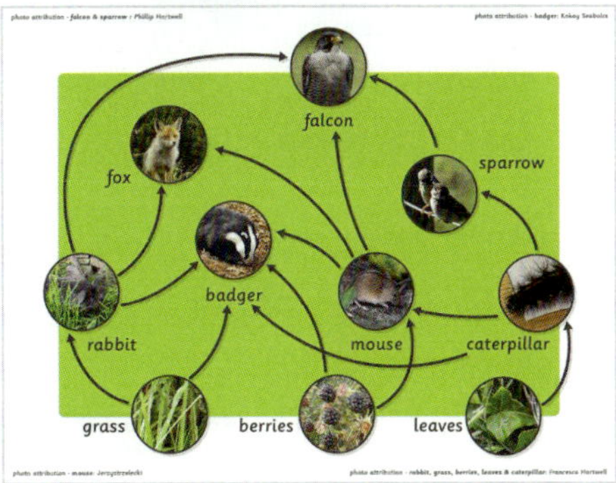

21. A food web shows how different food chains in a habitat link together. How many top predators (or top consumers) are shown in this woodland food web?

- **3**
- 6
- 7

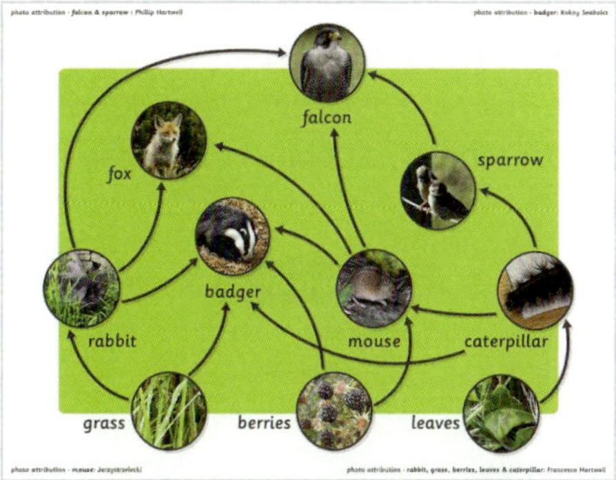

22. A food web shows how food chains link together in a habitat. If all of the badgers were removed, what might happen to the other animals and plants in this woodland food web?

1/3

- **There might be more rabbits and mice because they are prey for badgers.**
- There might be fewer foxes because they eat the same foods as badgers.
- The woodland might get overgrown with berry bushes without the badgers there to eat them.

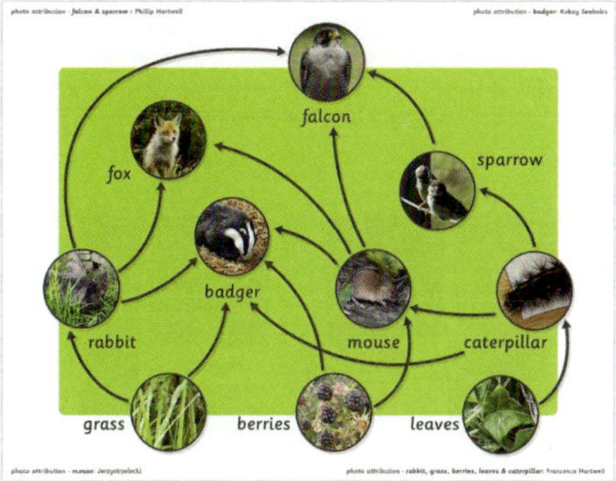

Pond Food Chains

Objective: I can interpret a variety of food chains, identifying producers, predators and prey.

Quick Search Ref: 10546

Level 1: Knowledge: I can use food chains to identify producers, predators and prey in a pond habitat.

Required: 7/10 | **Pupil Navigation:** on | **Randomised:** off

1. Algae are tiny green organisms which live in water. They produce their own food through photosynthesis, as plants do. Algae is a __________.

1/4

- predator
- prey
- **producer**
- consumer

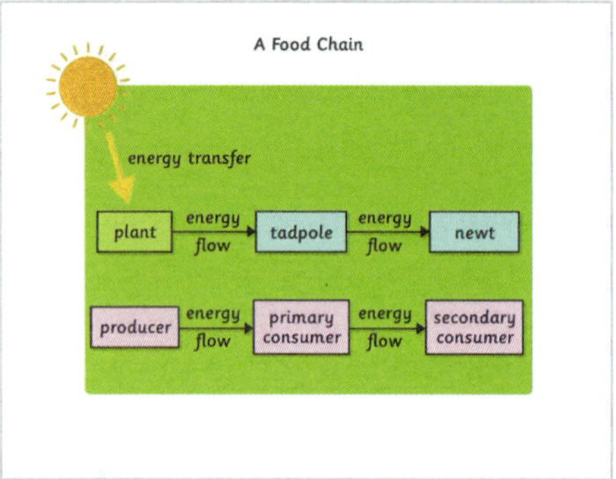

2. A predator is . . .

1/4

- an animal that eats animals that have been killed by other animals.
- an organism that can make its own food through photosynthesis.
- **an animal that kills and eats other animals.**
- an animal that is killed and eaten by other animals.

frog by H. Krisp, *kingfisher* by Ravi Vaidyanathan, *great diving beetle* by Evanherk

3. Which organisms are the predators in this food chain?

2/4

- **heron**
- algae
- **frog**
- water boatman

heron by Phillip Hartwell, *frog* by H. Krisp, *algae* by Tomwsulcer

4. What is the prey for the pond snail in this food chain?

1/4

- frog
- algae
- both of these
- **neither of these**

heron by Phillip Hartwell, *frog* by H. Krisp, *algae* by Tomwsulcer

5. Which animal is prey as well as predator in this food chain?

1/4

- pond weed
- pond snail
- **great diving beetle**
- heron

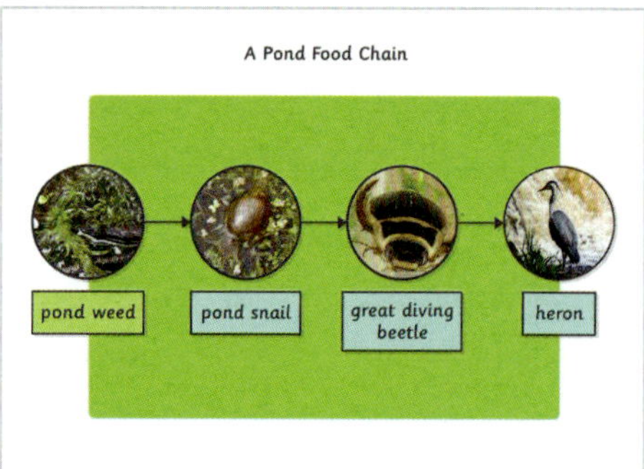

great diving beetle by Evanherk. *heron* by Phillip Hartwell

Level 1 *continued*

6. A food chain shows how _ _ _ _ _ _ is transferred between living organisms.

a b c

- energie
- **energy**
- energi
- energe

A Pond Food Chain

heron by Phillip Hartwell, *frog* by H. Krisp, *algae* by Tomwsulcer

7. What is the top predator in this food chain?

1/4

- water flea
- **kingfisher**
- frog
- algae

A Pond Food Chain

frog by H. Krisp, *kingfisher* by Ravi Vaidyanathan, *algae* by Tomwsulcer, *water flea* by Ayacop

8. What do the arrows between organisms in a food chain mean?

1/5

- kills
- eats
- is killed by
- is eaten by
- **energy flow**

A Pond Food Chain

heron by Phillip Hartwell, *frog* by H. Krisp, *algae* by Tomwsulcer

9. What is the producer in this food chain?

1/5

- the Sun
- pond snail
- heron
- **algae**
- frog

A Pond Food Chain

heron by Phillip Hartwell, *frog* by H. Krisp, *algae* by Tomwsulcer

10. Which animals are the prey animals in this food chain?

2/4

- algae
- **frog**
- heron
- **pond snail**

A Pond Food Chain

heron by Phillip Hartwell, *frog* by H. Krisp, *algae* by Tomwsulcer

Level 2: Understanding: I can create food chains to identify producers, prey and predators in a pond habitat.

Required: 3/3 **Pupil Navigation:** on
Randomised: off

11. Which of these pond organisms are producers?

2/6

- frog
- kingfisher
- **pond weed**
- heron
- **algae**
- great diving beetle

frog by H. Krisp, *kingfisher* by Ravi Vaidyanathan, *great diving beetle* by Evanherk, *algae* by Tomwsulcer.

Level 2 *continued*

12. Animals in a food chain are all _________.

1/4

- scavengers
- **consumers**
- carnivores
- producers

frog by H. Krisp, *kingfisher* by Ravi Vaidyanathan, *great diving beetle* by Evanherk, *algae* by Tomwsulcer,

13. Which arrows would you use to complete the food chain?

1/4

- **a**
- b
- c
- d

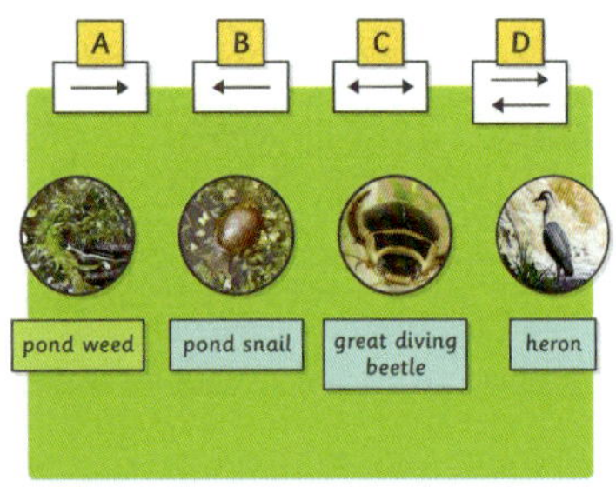

great diving beetle by Evanherk

Level 3: Extended reasoning about food chains: I can apply my knowledge and understanding of pond food chains to the world around me.

Required: 4/4 **Pupil Navigation:** on

Randomised: off

14. A food web shows how food chains link together in a habitat. Select **two** animals in this food web that are prey.

2/7

- heron
- otter
- **stickleback**
- **great diving beetle**
- algae
- **water flea**
- **pond snail**

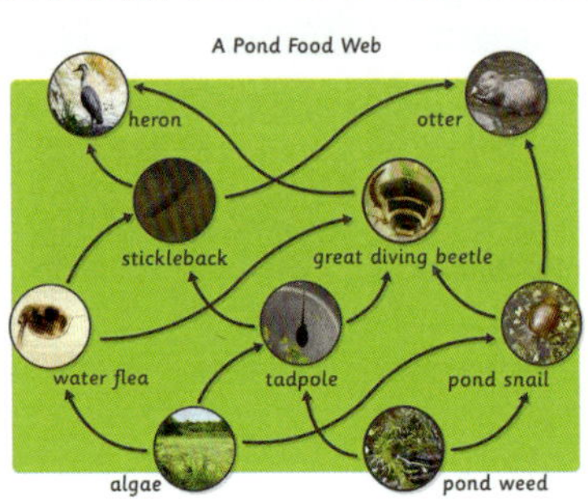

frog by H. Krisp, *kingfisher* by Ravi Vaidyanathan, *great diving beetle* by Evanherk, *water flea* by Ayacop, *algae* by Tomwsulcer, *stickleback* by S. Rae, *otter* by A. Amey

15. A food web shows how food chains link together in a habitat. How many top predators are shown in this pond food web?

1
2
3

- **2**
- 4
- 7

frog by H. Krisp, *kingfisher* by Ravi Vaidyanathan, *great diving beetle* by Evanherk, *water flea* by Ayacop, *algae* by Tomwsulcer, *stickleback* by S. Rae, *otter* by A. Amey

16. Mrs Coleman is building a pond in her garden. She wants to put in pond snails to stop the algae making the pond green. Use the pond food web to find out which predator she **doesn't want** in her pond.

1/4

- stickleback
- **great diving beetle**
- heron
- otter

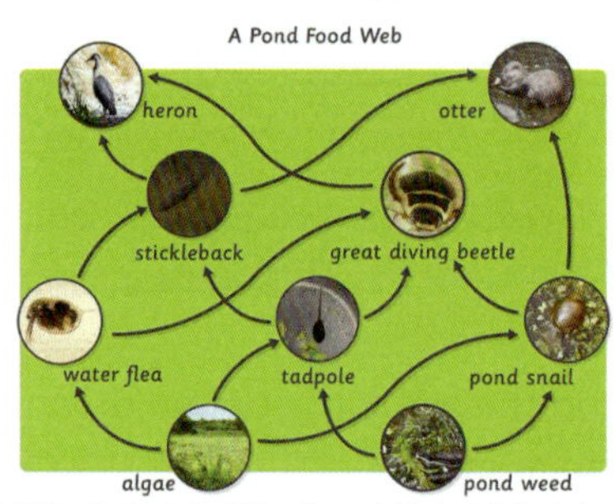

frog by H. Krisp, *kingfisher* by Ravi Vaidyanathan, *great diving beetle* by Evanherk, *water flea* by Ayacop, *algae* by Tomwsulcer, *stickleback* by S. Rae, *otter* by A. Amey

17. This food web shows the organisms that live in Ruby's pond.
Ruby knows that dragonfly nymphs eat insects and tadpoles. Would her pond provide food for adult dragonflies and their nymphs?

a
b
c

- Open question, no set answer

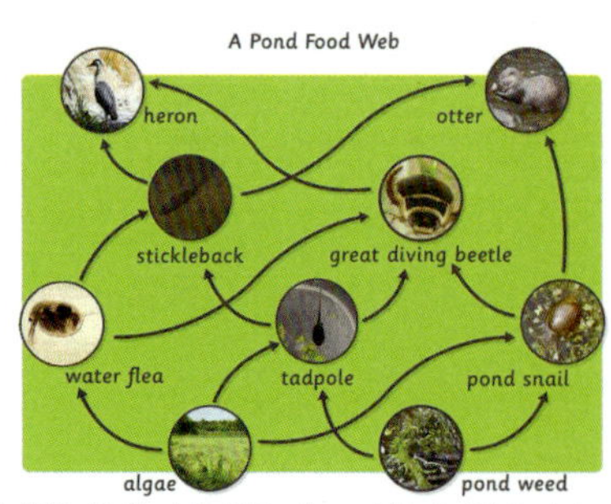

frog by H. Krisp, *kingfisher* by Ravi Vaidyanathan, *great diving beetle* by Evanherk, *water flea* by Ayacop, *algae* by Tomwsulcer, *stickleback* by S. Rae, *otter* by A. Amey

Classification Keys and Grouping Organisms

Objective: I know that living things can be grouped in a variety of ways.

Quick Search Ref: 10496

Level 1: Knowledge: I can group living things in different ways.

Required: 7/10 | Pupil Navigation: on | Randomised: off

1. Anything that is living is called . . .

- an animal.
- a plant.
- **an organism.**
- a human.

1/4

2. Which **3** of these things can all organisms do?

- **grow**
- see
- make noise
- **reproduce**
- **respire**

3/5

3. Which **3** of these things are organisms?

- **ant**
- rock
- **human**
- ruler
- **grass**

3/5

rock by Phillip Hartwell

4. Which **4** of these organisms are animals?

- **butterfly**
- **cat**
- **human**
- **earthworm**
- daffodil

4/5

cat by Phillip Hartwell; *earthworm* by Dodo-Bird

5. Which **4** of these organisms are plants?

- **oak tree**
- **tulip**
- koala
- **horse chestnut tree**
- **moss**

4/5

tulip by Phillip Hartwell

6. Which **2** of these pond organisms are invertebrates?

- duck
- **pond snail**
- **water flea**
- pond weed
- newt

2/5

snail by NOAA photo library; *frog* by H. Krisp; *water flea* by Ayacop

7. Which **3** of these animals are herbivores?

- **caterpillar**
- human
- lion
- **tortoise**
- **cow**

3/5

lion by Phillip Hartwell; *caterpillar* by Sherseydc

Level 1 *continued*

8. Which **4** of these organisms are animals?

4/5

- **dolphin**
- **sea star**
- seaweed
- **squid**
- **whale**

dolphin by Phillip Hartwell; *sea star* by Samuel Chow; *seaweed* by Raiyan; *squid* by NOAA/MBARI; *whale* by NOAA photo library

9. Which **2** of these rainforest animals are carnivores?

2/4

- **jaguar**
- sloth
- **crocodile**
- butterfly

sloth by Stefan Laube; *jaguar* by Phillip Hartwell; *crocodile* by Tomascastelazo

10. Which **2** of these animals are vertebrates?

2/5

- earthworm
- **human**
- **lizard**
- butterfly
- jellyfish

lizard by Stefan Laube; *earthworm* by Dodo-Bird

Level 2: Understanding: I can use classification keys to group, identify and name living things.

Required: 6/6 **Pupil Navigation:** on

Randomised: off

11. Which group of organisms eats other living things as food?

1/2

- **animals**
- plants

12. Plants make their own food by photosynthesis. Photosynthesis happens in the ______ of the plant.

a b c

- leafs
- **leaves**
- leeves
- leef
- leev
- leefs
- leaf
- leevs
- leafes
- leefes

13. Kiran, Lucy and Jamie are discussing whether or not a book is an organism. Who do you agree with?

1/3

- Kiran
- Lucy
- **Jamie**

14. Use the classification key to identify the organism.

1/5

- chestnut tree
- dolphin
- moss
- **pheasant**
- zebra

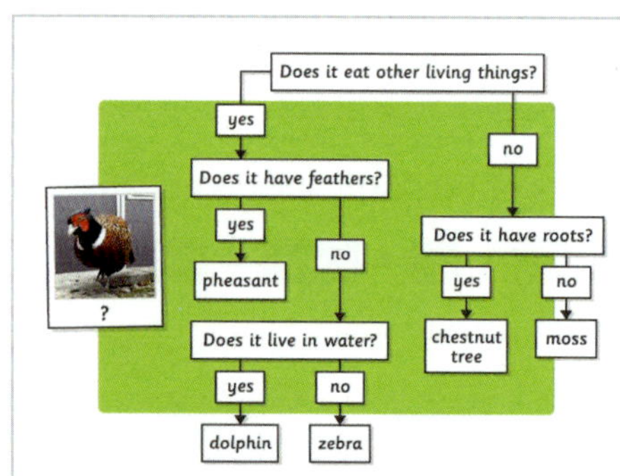

15. Use the classification key to identify the plant.

1/5

- **azalea**
- bluebell
- apple tree
- pine tree
- poppy

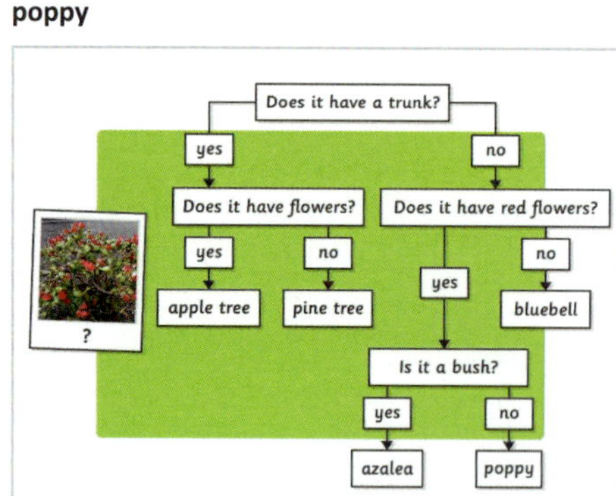

Level 2 *continued*

16. a b c This is a circular key. Start in the middle and follow the features outwards to identify the animal in the picture.

The animal has six legs - it is a ________.

- moth

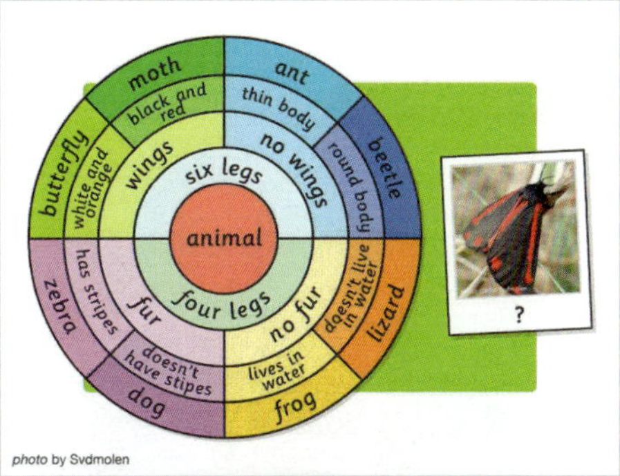

photo by Svdmolen

Level 3: application: I can apply my knowledge and understanding of grouping organisms to the world around me.

Required: 2/4 **Pupil Navigation:** on
Randomised: off

17. We classify organisms using **observable features**. An observable feature is . . .

1/3

- something we know.
- something we can see.
- something we guess.

18. Which of these questions could you use to classify the organisms in this picture?

1/4

- Does it have green leaves?
- Is it a plant?
- Is it pretty?
- Does it have flowers?

dandelion by Phillip Hartwell; *pine tree* by Dave Powell (USDA Forest Service)

19. What is a species?

1/3

- A group of organisms that look the same.
- A group of organisms that can reproduce together.
- A group of organisms that live together.

20. A sea cucumber is . . .

1/3

- a type of cucumber that grows in the sea.
- an animal that looks like a cucumber.
- an animal that eats cucumber.

Classification Vocabulary

Objective: I can use vocabulary relating to classification.

Quick Search Ref: 10550

Level 1: Knowledge: I can spell and define scientific words for classification.

Required: 9/9 **Pupil Navigation:** on **Randomised:** off

1. Scientists sort things into groups so that they can be identified. We call this ___________.

1/4

- sorting
- **classification**
- grouping
- organising

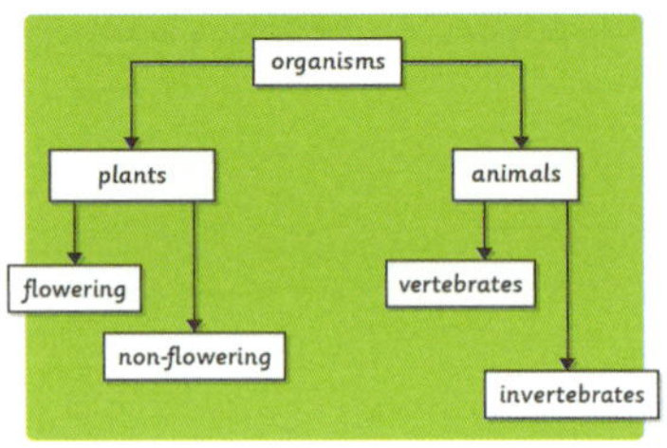

2. A classification _ _ _ uses questions to group organisms by their features.

- kee
- **key**
- quay

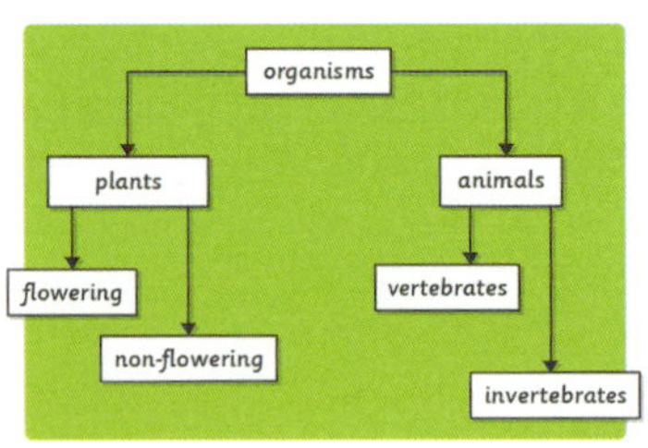

3. An organism is a . . .

1/4

- plant or animal.
- non-living thing.
- **living thing.**
- living thing that isn't a plant or animal.

4. An organism that used to be alive but isn't anymore is _ _ _ _.

- **dead**
- ded.
- **dead.**
- ded

5. A volcano, a plastic bottle and a gold bracelet are things that . . .

1/3

- are alive.
- used to be alive, but are now dead.
- **have never been alive.**

6. A plant . . .

1/3

- **can make its own food.**
- is always green.
- always grow out of the ground.

7. A human, a salamander and an ant are all _ _ _ _ _ _ _.

- animul
- animuls.
- **animal.**
- animl.
- **animal**
- animul.
- **animals.**
- **animals**
- animl
- organism
- animuls
- animls.
- organisms
- animls

Level 1 *continued*

8. An animal with an internal skeleton is called . . .

1/3

- a vertebrate.
- an invertebrate.
- a human.

9. An invertebrate is . . .

1/5

- an animal with an internal skeleton.
- an animal without an internal skeleton.
- an insect.
- a small animal.
- an animal that lives in water.

Investigating Leaf Litter Using Identification Keys - Plan and Do

Objective: **Students will plan a survey investigation to identify the most common type of invertebrate found in leaf litter in their local environment. They will need to be provided with recording and investigation materials.**

Quick Search Ref: 10563

Level 1: Planning an investigation to survey what organisms live in leaf litter: I can ask relevant scientific questions; I can make a prediction with a reason.

Required: 7/7 **Pupil Navigation:** on **Randomised:** off

1. These children have been discussing which invertebrates might live in leaf litter in their local area. What could they investigate to find out?

1/3

- What different types of invertebrates live in the local area?
- What is the most common type of invertebrate in the local area?
- Which section of the local area has the most invertebrates?

2. Working safely: Select the most important safety rule for working with animals.

1/3

- **Wear gloves to protect your hands.**
- **Wash your hands afterwards.**
- **Wear goggles to protect your eyes.**

3. You are going to collect some leaf litter to investigate the invertebrates that live in it. What will you **look for**?

1/3

- **types of animals**
- **types of plants**
- **types of invertebrates**

4. You are going to collect some leaf litter to investigate the invertebrates that live in it. What will you **count**?

1/3

- **number of types of invertebrates**
- **number of legs**
- **number of each type of invertebrate**

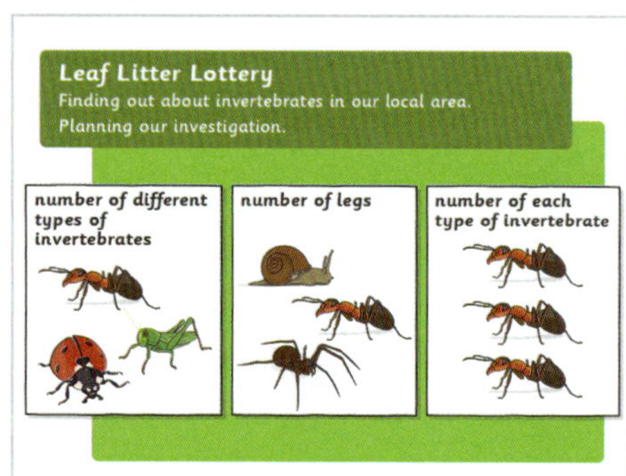

Level 1 *continued*

5. What scientific question are you going to investigate?

1/4

- **What is the most common type of invertebrate living in leaf litter in our local area?**
- How many types of invertebrates are there in our local area?
- What is the most common animal in our local area?
- What colour are the invertebrates in our local area?

6. What equipment will you need for your investigation?

1/6

- containers with mesh lids
- magnifying glass
- pooters
- small brushes
- identification key
- **all of these things**

7. What do you think will be the most common invertebrate in the leaf litter in your local area? Why do you think this?

a b c

- Open question, no set answer

Level 2: Measuring and recording information from a survey: I can make careful and accurate observations; I can record data.

Required: 7/7 **Pupil Navigation:** on

Randomised: off

8. Before you start your survey, you need to draw a table to record your results. What should the titles of the two columns be?

2/5

- colour of invertebrate
- **type of invertebrate**
- where the invertebrate was found
- how many legs the invertebrate has
- **number of invertebrates**

9. How will you know what each invertebrate is?

1/3

- **use an identification key**
- guess
- ask a friend

Level 2 *continued*

10. Before you record the number of each type of invertebrate, what else do you need to record in your table?

1/4

- the number of legs of each invertebrate has
- the colour of each invertebrate
- **the name of each type of invertebrate you find**
- where you found each invertebrate

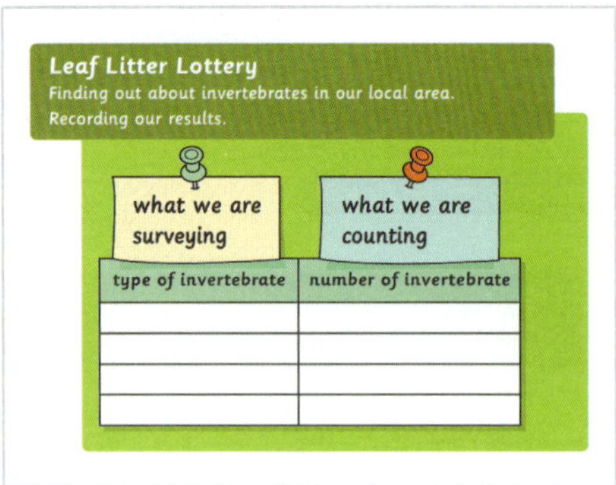

11. How will you record the number of each type of invertebrate in the table?

1/3

- by using a ruler
- **by keeping a tally**
- by writing the number once you have counted all of the invertebrates

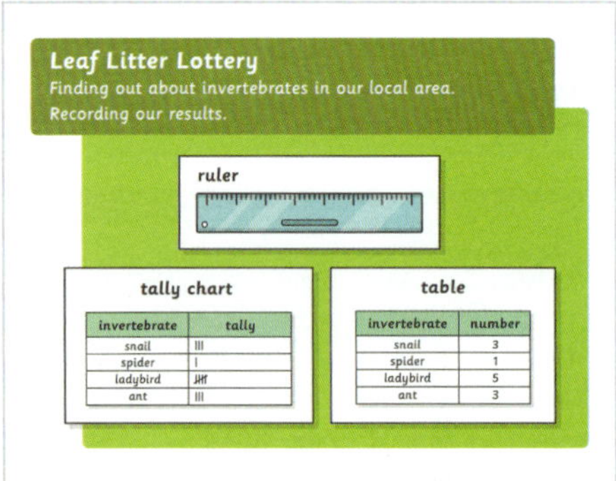

12. Why should you keep each type of invertebrate in a separate container?

1/3

- **If you put all of the invertebrates in one container, they might eat each other.**
- So they have space to move around.
- So that you don't get confused.

13. What must you do after you have finished the survey?

1/4

- wash your hands
- clean the equipment you have used
- return the invertebrates to their habitat
- **all of the above**

14. What other things do you need to consider before starting your investigation?

a b c

- Open question, no set answer

Investigating Leaf Litter using Identification Keys - Review

Objective: Students will review results from a survey investigation to identify the most common type of invertebrate found in leaf litter in their local environment. Recording and investigation materials used in 'Investigating Leaf Litter using Classification Keys - Plan and Do' will need to be provided.

Quick Search Ref: 10564

Level 1: Reviewing the plan for an investigation to survey what organisms live in leaf litter: I can ask relevant scientific questions; I can make a prediction with a reason.

Required: 5/5 **Pupil Navigation:** on **Randomised:** off

1. You carried out an investigation to find out which invertebrates live in your local area. What did you investigate?

1/3

- types of invertebrates in the local area
- most common invertebrate in the local area
- the section of the local area with the most invertebrates

2. You collected some leaf litter to investigate invertebrates. What were you **looking for**?

1/3

- types of animals
- types of plants
- types of invertebrates

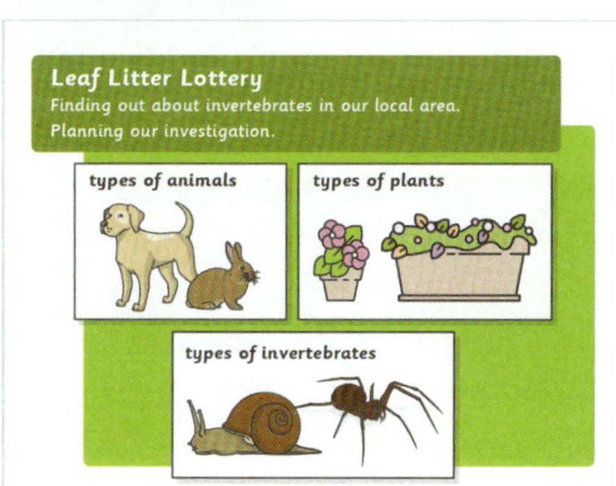

3. You collected some leaf litter to investigate invertebrates. What did you **count**?

1/3

- number of different types of invertebrates
- number of each invertebrate
- number of legs

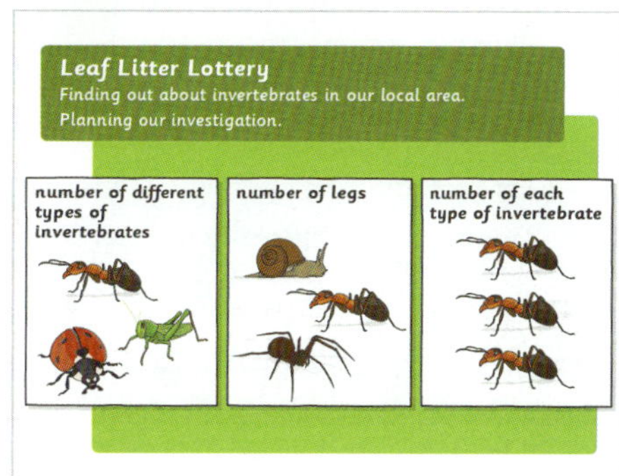

4. Which scientific question did you investigate?

1/4

- What is the most common invertebrate in our local area?
- What is the most common animal in our local area?
- What colour are the invertebrates in our local area?
- How many types of invertebrates are there in our local area?

Level 1 *continued*

5. Before starting your survey, you drew a results table. What are the titles of the two columns?

2/5

- **type of invertebrate** ✓
- colour of invertebrate
- number of legs
- **number of invertebrate** ✓
- where the invertebrate was found

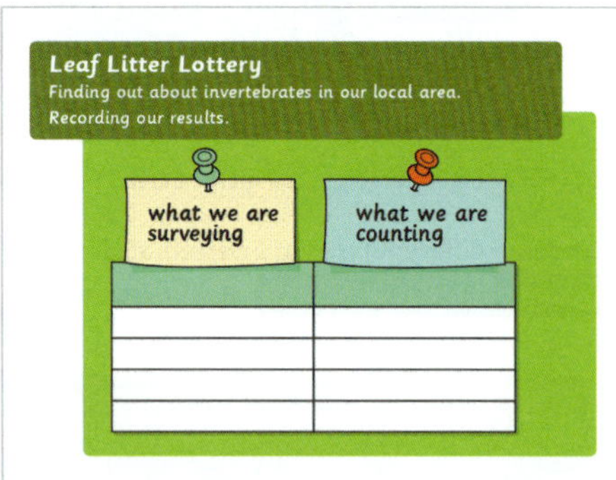

Level 2: Presenting data from a survey in a bar chart: I can present data in a bar chart.

Required: 5/5 **Pupil Navigation:** on

Randomised: off

6. What is the best way to show the results of your survey?

1/3

- in a tally chart
- **in a bar chart** ✓
- in a pictogram

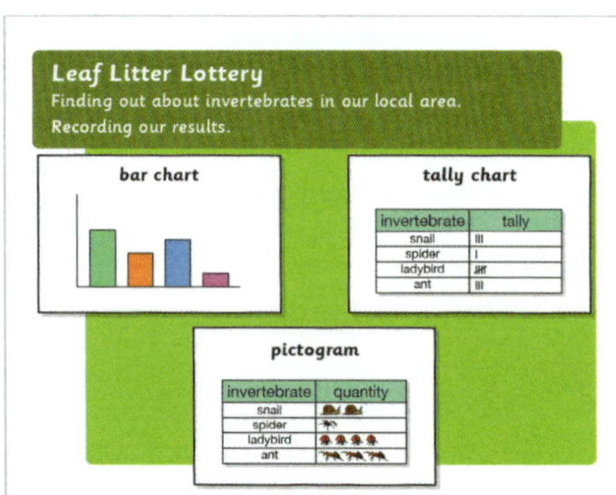

7. Which label will you put along the horizontal axis?

1/2

- **type of invertebrate** ✓
- number of invertebrates

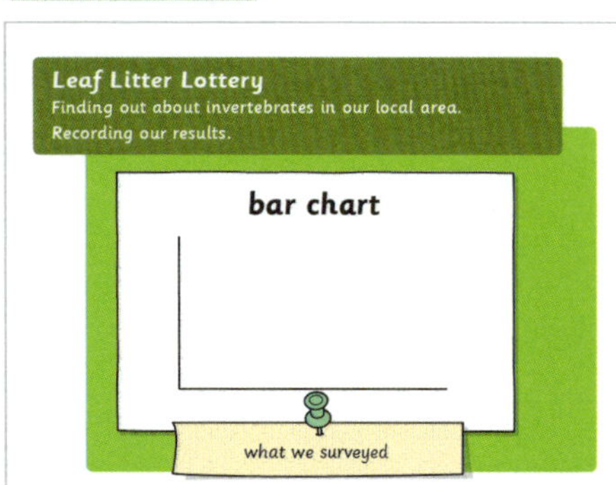

8. Which label will you put along the vertical axis?

1/2

- type of invertebrate
- **number of invertebrate** ✓

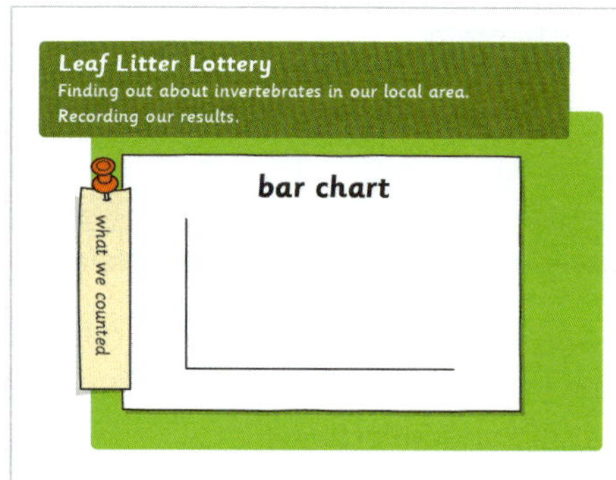

9. How many bars will you plot on your bar chart?

- Open question, no set answer

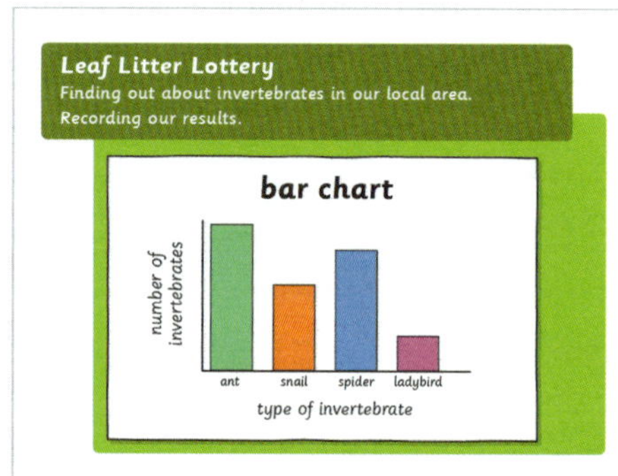

10. Complete the title for your graph: A graph to show . . .

1/2

- **the number of different types of invertebrates found in the local area.** ✓
- the different types of invertebrates found in the local area.

Level 3: Analysing and reviewing: I can interpret results; I can draw conclusions and suggest improvements.

Required: 6/6 **Pupil Navigation:** on

Randomised: off

11. What does the tallest bar on your bar chart show you?

1/3

- **which type of invertebrate was most common** ✓
- how many types of invertebrates there were altogether
- which type of invertebrate was the least common

12. What does the lowest bar on your bar chart show you?

1/3

- how many types of invertebrates there were altogether
- **which type of invertebrate was the least common** ✓
- which type of invertebrate was most common

Level 3 *continued*

13. What does the number of bars on your bar chart show you?

1/3

- which type of invertebrate was most common
- which type of invertebrate was the least common
- how many types of invertebrates there were

14. Did your results agree with your prediction? Explain how you know.

a b c

- Open question, no set answer

15. Would you do anything differently if you did this investigation again?

a b c

- Open question, no set answer

16. Do you think that your results would be the same if you did this investigation in a pond habitat? Explain your answer.

a b c

- Open question, no set answer

Environmental Changes in Habitats

Objective: I can recognise that environments can change and that this can sometimes pose dangers to living things.

Quick Search Ref: 10608

Level 1: Knowledge: I can describe environmental changes.

Required: 7/10 **Pupil Navigation:** on **Randomised:** off

1. Which of these are natural causes of change to the environment?

3/5

- **storms**
- **volcano eruptions**
- deforestation
- pollution
- **droughts**

drought by Axel Kristinsson; pollution by Julien Gomba

2. Deforestation is when . . .

1/3

- a tree is cut down in a forest.
- the trees in a forest are blown down in a storm.
- **large areas of forest are cut down.**

3. P_ _ _ _ _ _ _ _ is something introduced into an environment that has harmful effects.

- **pollution**
- polution

oil spill by Julien Gomba

4. A nature reserve is . . .

1/3

- an area of land that is beautiful.
- **an area of land which is looked after to protect the plants and animals that live there.**
- a place where animals are kept in enclosures to keep them them safe.

nature reserve by Chris Mckenna

5. Littering is a kind of . . .

1/3

- **pollution.**
- urban development.
- deforestation.

nature reserve by Chris Mckenna

6. Which of these are man-made changes to the environment?

3/5

- **introducing invasive species**
- volcano eruptions
- **deforestation**
- **pollution**
- droughts

drought by Axel Kristinsson; pollution by Julien Gomba

Level 1 *continued*

7. Over 200 years ago, explorers brought back rhododendron plants to Britain. These plants have spread out of gardens and now can be found in woods around the country. Which type of environmental change is this?

1/3

- introduction of invasive species
- deforestation
- intensive farming

rhododendron

rhododendron by A Amey

8. Which of these changes to the environment make better habitats for plants and animals?

3/5

- leaving areas of grass unmown
- filling in a pond and building a playground
- tidying away all fallen leaves and weeds
- planting brightly coloured, sweet smelling flowers
- building a wildlife bridge over a road

garden by Francesca Hartwell

9. Where does pollution happen?

1/6

- in air
- in water
- on land
- in space
- in all of these places
- in none of these places

oil spill by Julien Gomba

10. Gases that trap heat inside the Earth's atmosphere are called . . .

1/3

- poisonous gases.
- laughing gases.
- greenhouse gases.

Level 2: I can describe how changes to an environment could help or endanger living things.

Required: 6/6 **Pupil Navigation:** on
Randomised: off

11. Choose two reasons why areas of sea need to be looked after in the same way as nature reserves protect areas of land.

2/3

- Over fishing has led to some fish species becoming endagered.
- People like to go diving in the sea to look at the fish and coral.
- Damage to the bottom of the sea is harming the plants and animals that grow and live there.

damaged coral

damaged coral by US Geological Survey

12. How does deforestation affect habitats?

2/4

- There are fewer kinds of plants because the animals that spread the seeds move away or die.
- The land that is farmed after the trees have been cleared stays healthy and fertile.
- Animals such as orangutans, which live in the trees, lose their food and shelter.
- The water supply outside the cleared area gets better.

deforestation

Level 2 *continued*

13. Hedgehogs travel about one mile every night to find food and a mate. The number of hedgehogs is getting smaller because gardens are surrounded by walls and fences, which makes it difficult for hedgehogs to move around. Which change would help hedgehogs to feed and breed?

1/4

- taking down all of the garden fences
- putting out food for hedgehogs
- making a small hole in the bottom of garden fences and walls.
- building a hedgehog house in the garden

hedgehog by Phillip Hartwell

14. What are some of the ways that bees and butterflies are harmed by intensive farming?

1/4

- Some pesticides, which farmers use to protect their crops, are harmful to many kinds of insects, not just the ones that eat their plants.
- Farmers can produce more food on smaller areas of land if they farm intensively.
- Intensive farming can make food cheaper for people to buy in the supermarket.
- When farmers clear fields of wildflowers and grasses to grow one kind of plant such as wheat, they take away the food and shelter that many insects need.

15. Which of these are examples of climate change?

1/4

- In Britain, there was a very hot and dry summer in 1976. Rivers dried up and water supplies started to run out.
- In Papua New Guinea, people used to know when the the wet season would start, but now they don't.
- In Pakistan, river flooding happens more often so people have to move every year instead of staying in one place for 3 or 4 years.
- In the United States, the Mississippi River flooded in 1927 destroying the homes of thousands of people.

16. Some people don't think dropping cigarette ends is littering. How do cigarette ends harm habitats?

1/5

- Harmful chemicals in cigarette ends pollute soil and water.
- Plastic fibres in cigarette ends don't break down as natural materials do.
- Animals such as fish and turtles eat cigarette ends.
- none of these
- all of these

Level 3: Working scientifically to investigate environmental changes. I can ask relevant scientific questions. I can draw conclusions.

Required: 4/4 **Pupil Navigation:** on
Randomised: off

17. The children at High Lane School want to attract more bees to their school garden by planting plants with colourful flowers. What question could they investigate to decide which plant to use?

1/3

- What will happen to the number of bees when we change the time of planting?
- What will happen to the number of bees when we change the colour of the flower?
- What will happen to the number of flowers when we change the type of plant?

Level 3 *continued*

18. The children in Mrs Dunne's class have been investigating the greenhouse effect. The children put two thermometers outside and covered one with a glass jar. They recorded the temperature on the thermometers every half hour and plotted their results on a line graph. Which conclusion do their results support?

1/3

- **Our results show that climate change causes global warming.**
- **Our results show that the air temperature inside the jar stayed the same as the temperature outside the jar.**
- **Our results show that the air temperature inside the jar got hotter than outside the jar.**

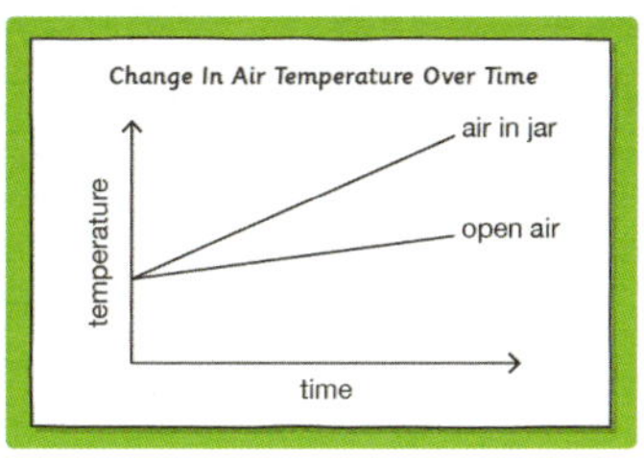

19. The table shows the number of birds that visited the High Lane School playground on the same date each year for four years. What environmental change might explain why more birds were seen in year 3?

1/4

- **The weather was better in the third year.**
- **Bird feeders were put up in the playground before the third year survey.**
- **Bird feeders were removed from the playground before the third year survey.**
- **Two cats came to live in the neighbourhood before the third year survey.**

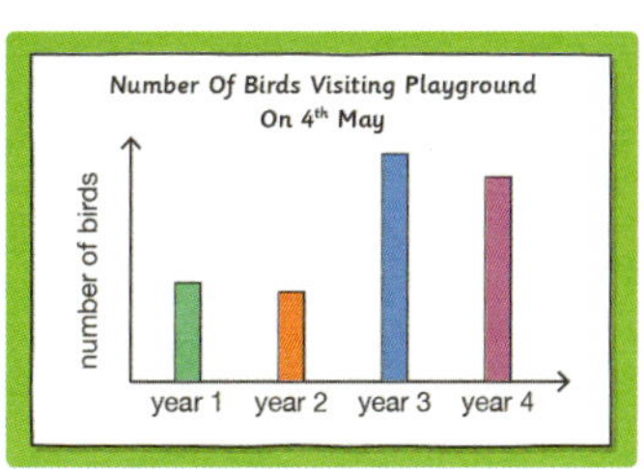

20. How could the children in Mrs Dunne's class find out how many types of animals they have attracted to their mini pond?

1/3

- **The children could research food chains in a pond habitat.**
- **The children could do a fair test investigation into whether tadpoles prefer to be in shade or light.**
- **The children could do a pond dipping survey investigation.**

Level 4: Extended reasoning about environmental changes in habitats. I can explain how people, weather and the environment can affect living things.

Required: 4/4 **Pupil Navigation:** on
Randomised: off

21. Use the orangutan fact file to identify why deforestation is so harmful to orangutans.

2/3

- **The orangutans can't move quickly through the trees, so hunters can catch them easily.**
- **Orangutans only have babies every eight years, so it takes a long time for their numbers to increase.**
- **The fruits which are the orangutans main food source don't grow any more.**

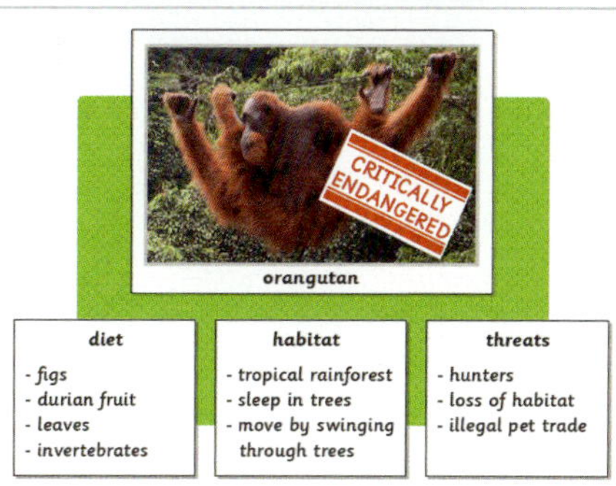

Level 4 *continued*

22. The children have been talking about the effects of environmental changes. Who do you think has the best understanding?

1/3

- Bella
- George
- Hafsah

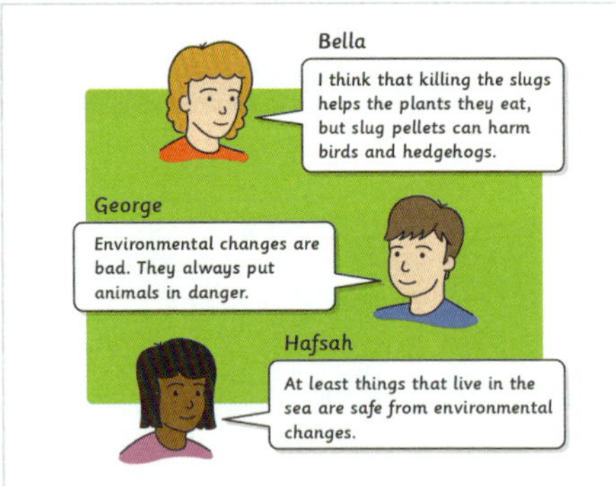

23. The debating club is holding a debate 'We believe that palm oil farming is a good thing'. Which of these statements are in favour of palm oil farming?

3/6

- Palm oil plantations are planted by clearing large areas of forest.
- Palm oil takes less land to produce more oil than soybean or coconut oil.
- Palm oil farming provides lots of jobs for people living in poor countries.
- There are reports that young children are working on palm oil plantations.
- Palm oil trees don't need many fertilisers or pesticides.
- Palm oil is high in saturated fats which is bad for people's health.

24. The Red List of Threatened Species lists all of the species we know about and how endangerd they are. Starting with 1) Extinct - none known to be living, arrange the conservation status definitions from most endangered to least endangered.

- none known to be living
- only living in protected areas
- extremely high danger of becoming extinct
- very high danger of becoming extinct
- high risk of becoming endangered
- likely to become endangered soon
- lowest risk, many living

Environmental Changes in Habitats Vocabulary

Objective: I recognise that environments can change and that this can sometimes pose dangers to living things.

Quick Search Ref: 10607

Level 1: Knowledge: I can spell and define scientific words for changes to habitats.

Required: 10/10 **Pupil Navigation:** on **Randomised:** off

1. What is the scientific term for a living thing?

abc

- **organisms**
- **organism**
- organisum
- organizm

monkey, plant by Phillip Hartwell

2. An organism's _______ provides food, shelter and protection.

abc

- **habitats**
- **habitat**
- habetat
- home

3. A ______ is something that might harm an animal or plant.

abc

- **threat**
- treat

4. The conditions that a plant or animal lives in are its . . .

1/3

- territory.
- climate.
- **environment.**

5. Deforestation is . . .

1/3

- **when large areas of forest are destroyed.**
- when a tree is cut down.
- when palm oil plantations are planted.

6. Some examples of _________ are: oil spils, littering, car exhaust gases, plastic in the sea.

abc

- polushun
- **pollution**
- pollushun
- polution

oil spill by Julien Gomba; sea litter by Vaidehi Shah; plastic litter by Ben Van der Ploeg

7. An area of natural habitat that is looked after to protect the plants and animals that live there is . . .

1/3

- an animal's territory.
- a zoo.
- **a nature reserve.**

west sedgemoor by Chris Mckenna

Level 1 *continued*

8. A _ _ _ _ _ _ _ report describes the wind, rain and sunshine for a place at a certain time.

a b c

- wethr
- wever
- weather
- weathr
- wether
- whether

9. An area's _ _ _ _ _ _ _ describes its weather patterns over many years.

1/3

- environment.
- climate.
- habitat.

10. When a species of animal or plant is extinct . . .

1/3

- there are so few left that scientists think they will soon die out.
- there are only a few of the species left living in protected areas.
- scientists think there are no longer any living.

Primary Science Y3/4

Matter

States of Matter

Evaporating and Condensing

Objective: I can describe evaporating and condensing in terms of changes in temperature.

Quick Search Ref: 10646

Level 1: Knowledge: I can describe the processes of evaporating and condensing.

Required: 7/10 **Pupil Navigation:** on **Randomised:** off

1. All substances are made up of ________.

1/3

- state.
- volume.
- **matter.**

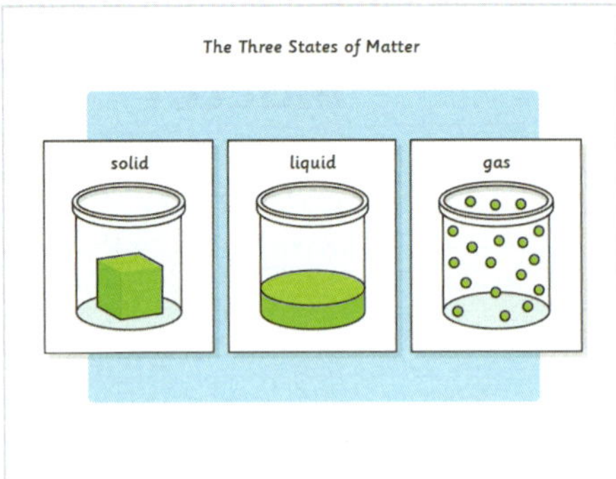

2. Solid, liquid and gas are all _ _ _ _ _ _ of matter.

- stayts
- **states**
- staytes
- state

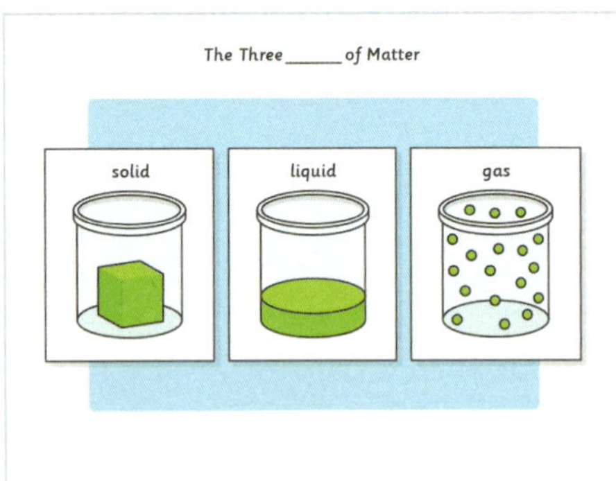

3. _ _ _ _ _ _ _ _ _ _ _ _ is the change of state from liquid to gas.

- **evaporation**
- **evaporating**
- evaporashun
- evapouration
- condensation
- freezing
- condensing
- evapourating
- melting

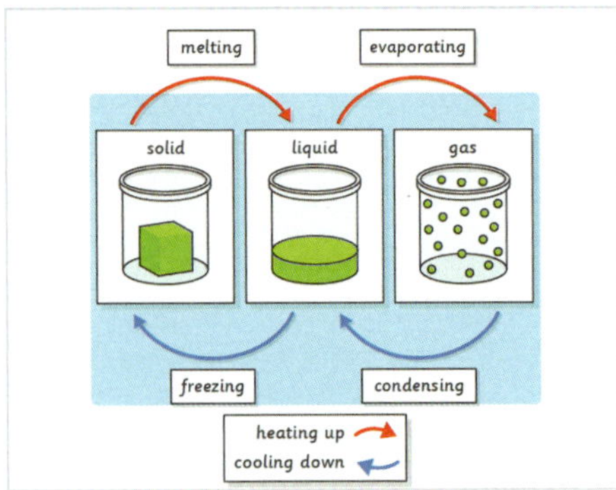

4. The opposite process to condensing is . . .

1/3

- melting
- freezing
- **evaporating**

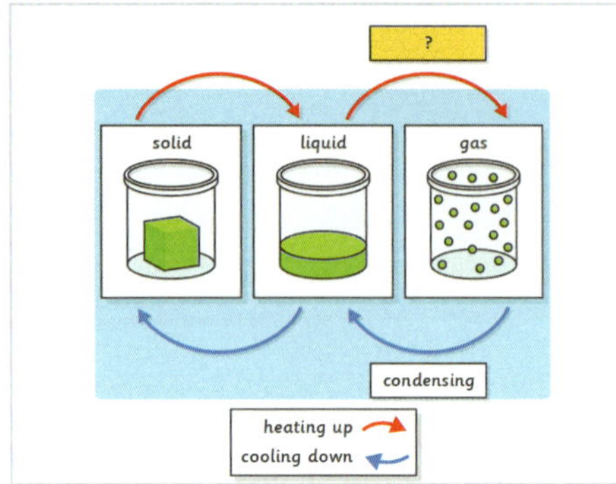

5. Which of these statements about evaporating and condensing isn't true?

1/4

- **A new material is formed when a substance evaporates.**
- Condensing is a reversible change.
- Water and water vapour are the same material in different states.
- Evaporation is a reversible change.

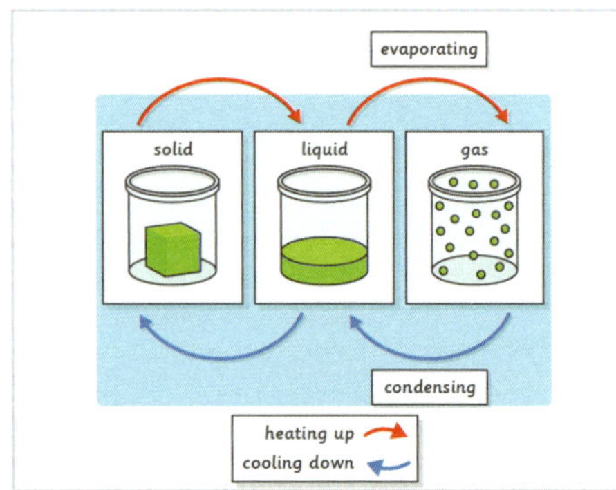

6. Condensing is the ____________ from gas to liquid.

1/4

- change of material
- change of temperature
- **change of state**
- change of shape

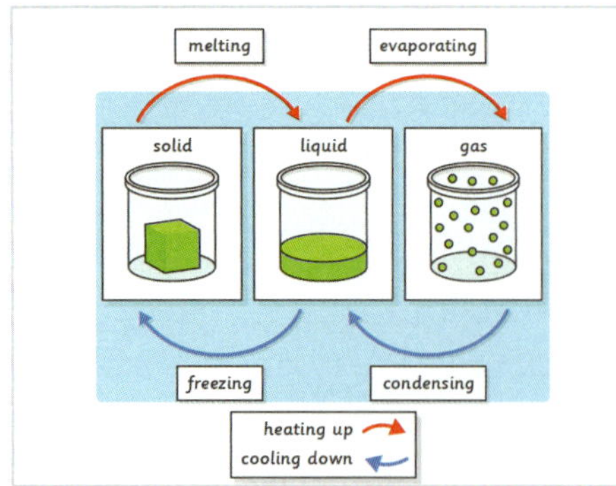

Level 1 *continued*

7. How can you tell that a puddle of liquid is evaporating?

1/3

- A puddle of liquid goes hard when it evaporates.
- Smoke comes off a puddle of liquid when it evaporates.
- A puddle of liquid gets smaller when it's evaporating.

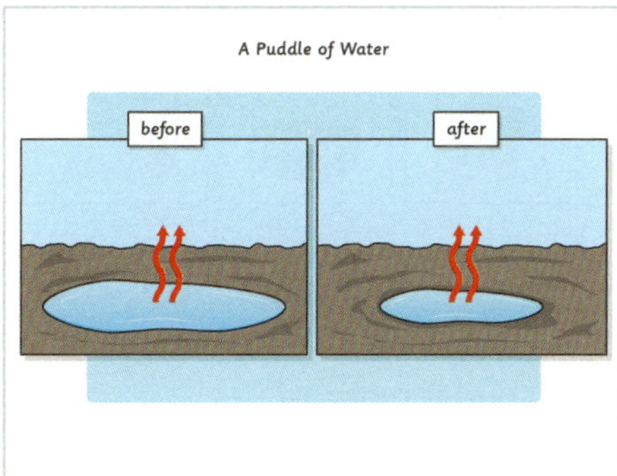

8. Once a gas has condensed to form a liquid, is it possible to change the material back into a gas?

1/3

- Yes. Evaporating the liquid will make the material change state back to a gas.
- No. Condensing is a change that can't be reversed.
- Yes, but only if the material is water.

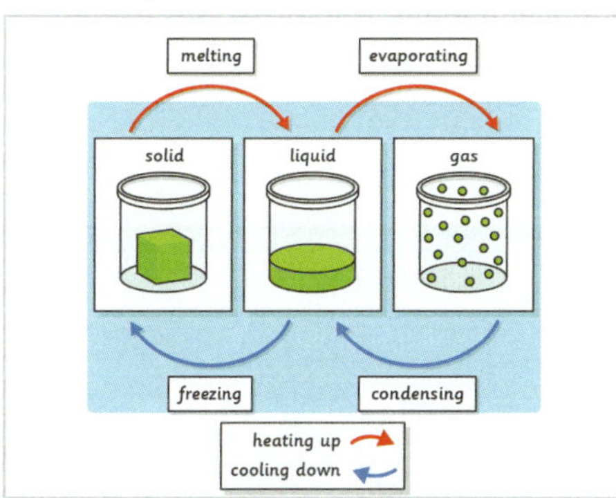

9. How can you tell that water vapour gas is starting to condense?

2/3

- You can see droplets of liquid water forming on a cold surface.
- You can see bubbles forming in boiling water.
- You can see a cloud of steam.

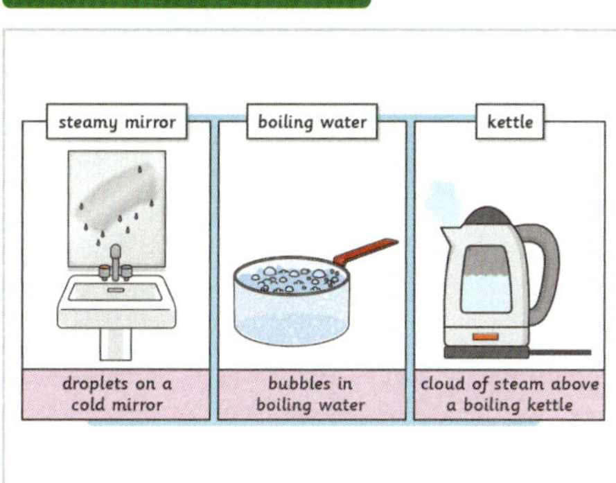

10. What is formed when oxygen gas condenses?

1/3

- liquid water
- liquid air
- liquid oxygen

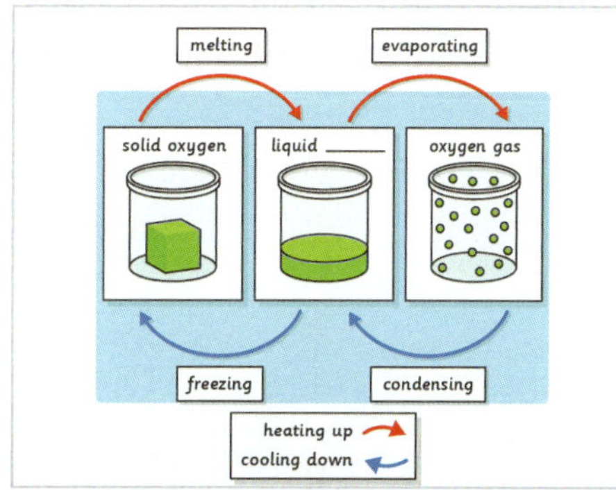

Level 2: Understanding: I can explain that changing temperatures can cause materials to evaporate or condense.

Required: 8/8 **Pupil Navigation:** on
Randomised: off

11. The ___________ of a substance is the measure of how hot it is.

a b c

- thermometer
- tempratur
- temperature
- temperatyur
- temprature
- Celsius
- degrees
- temprachure
- temperatur
- temperachure

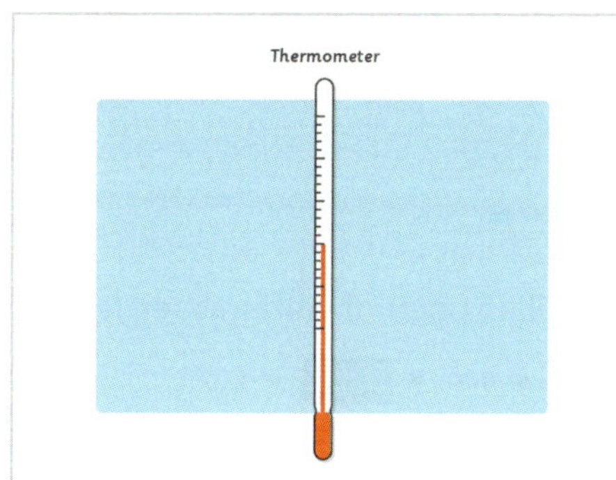

12. _______ a liquid causes it to evaporate more quickly.

1/3

- Cooling
- Freezing
- Heating

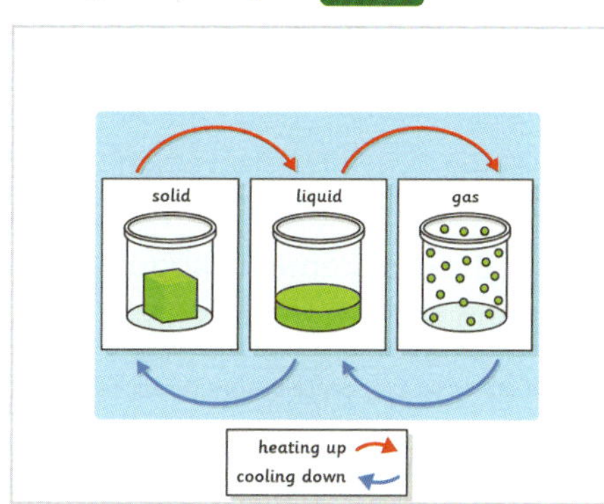

Level 2 *continued*

13. The temperature of something is measured in _ _ _ _ _ _ _ Celsius (°C).

a b c

- **degrees**
- **digrees**

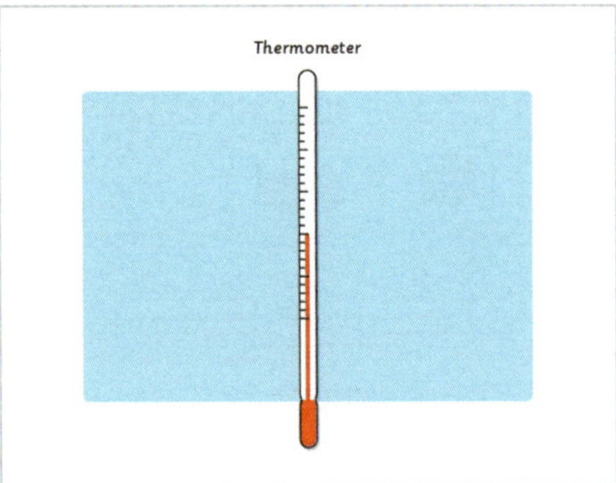

14. The highest temperature a liquid reaches before changing state to become a gas is the substance's . . .

1/3

- **boiling point**
- **freezing point**
- **melting point**

15. The boiling point of water at sea level is . . .

1/4

- **-10 °C**
- **0 °C**
- **100 °C**
- **110 °C**

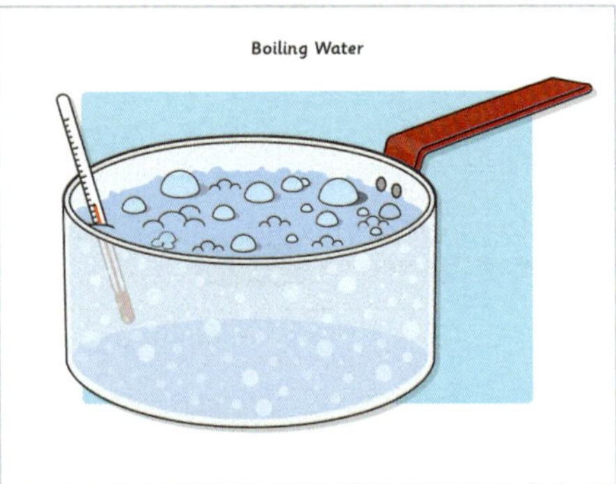

16. Which of these statements about evaporation is incorrect?

1/3

- **A liquid starts to evaporate before it reaches its boiling point.**
- **A liquid can only evaporate when it reaches its boiling point.**
- **Evaporation happens faster at higher temperatures.**

17. _______ a gas to its boiling point causes it to condense into a liquid.

1/3

- **Cooling**
- **Heating**
- **Dissolving**

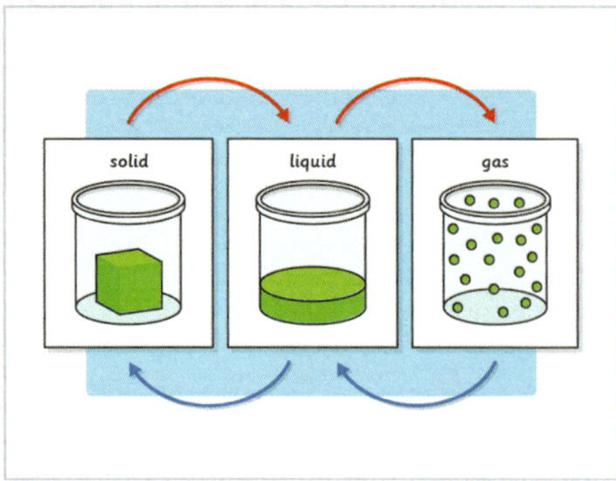

18. A rain cloud is made of . . .

1/2

- **water vapour gas that has evaporated from the sea.**
- **droplets of liquid water that have condensed from water vapour gas.**

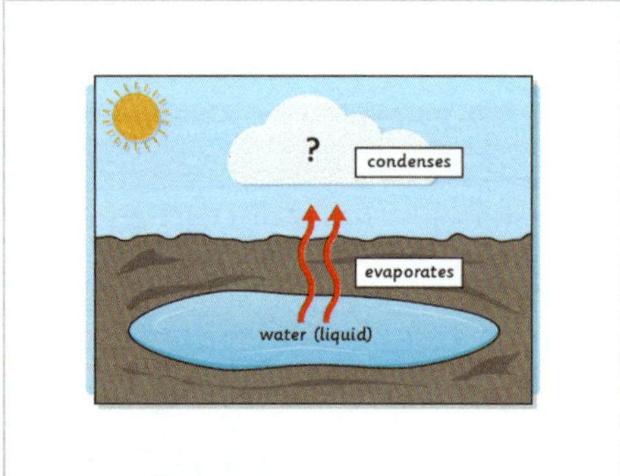

Level 3: Application: I can apply my knowledge and understanding of evaporating and condensing to the world around me.

Required: 2/2 **Pupil Navigation:** on

Randomised: off

19. Explain why dogs pant in hot weather.

a b c

- Open question, no set answer

Level 3 *continued*

20. What could Mr Brown do to make his washing dry faster?

a
b
c

- Open question, no set answer

Melting and Freezing

Objective: I can describe melting and freezing in terms of changes in temperature.

Quick Search Ref: 10647

Level 1: Knowledge: I can describe the processes of melting and freezing.

Required: 6/9 **Pupil Navigation:** on **Randomised:** off

1. Substances are made up of . . .

1/3

- state.
- volume.
- **matter.**

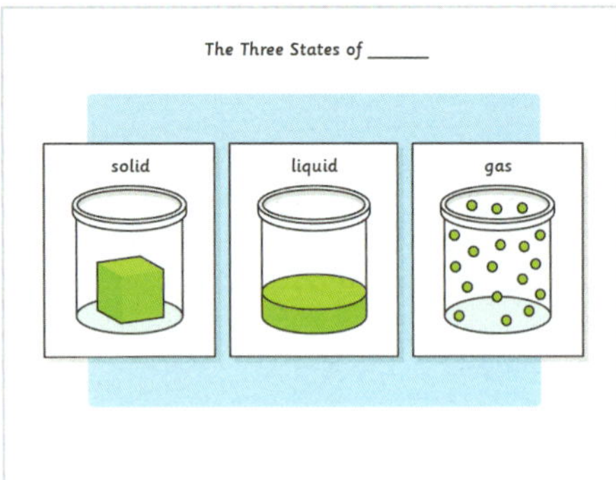

2. Solid, liquid and gas are all ______ of matter.

a b c

- staytes
- **states**
- state
- stayts

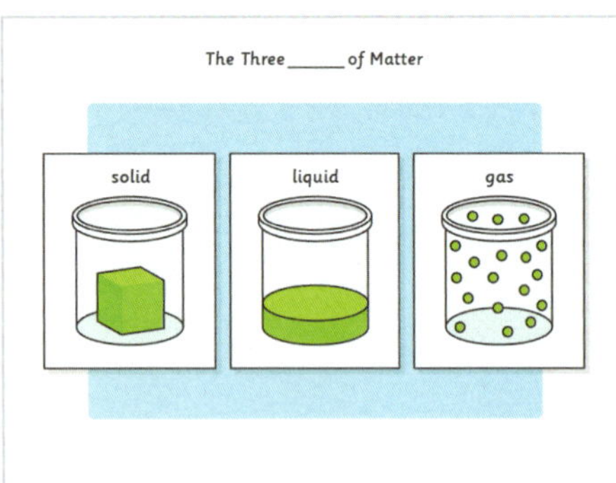

3. ________ is the change of state from liquid to solid.

a b c

- **freezing**
- frezing
- freazing
- melting
- evaporating
- condensing
- freesing

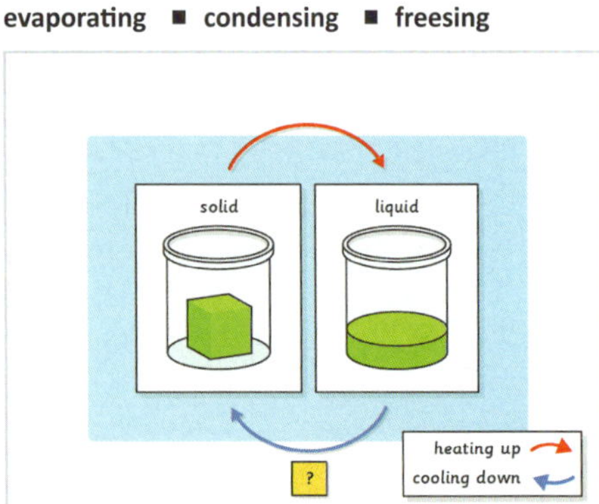

4. The opposite process to freezing is . .

1/3

- evaporating
- **melting**
- condensing

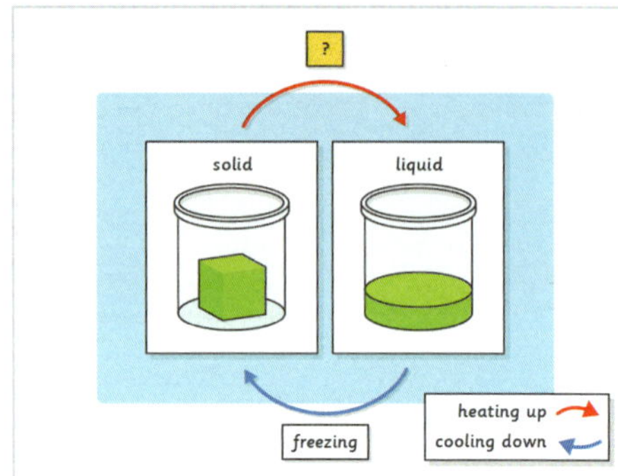

5. Which of these statements about melting and freezing is incorrect?

1/4

- **A new material is formed when a substance freezes.**
- Melting is a reversible change.
- Ice and water are the same material in different states.
- Freezing is a reversible change.

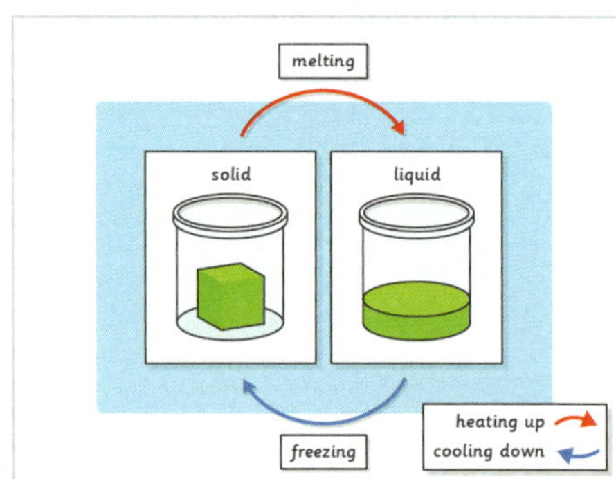

6. Melting is the ____________ from solid to liquid.

1/4

- change of material
- change of temperature
- **change of state**
- change of shape

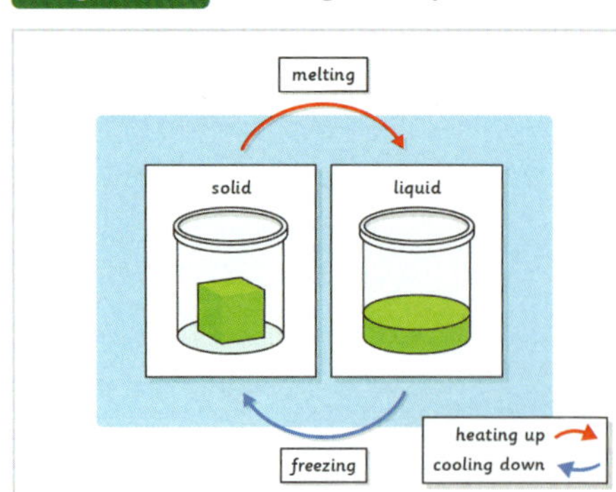

Level 1 *continued*

7. How can you tell that a solid is melting?

1/3

- A solid gets harder when it melts.
- Smoke comes off a solid when it melts.
- **Liquid starts to form when a solid melts.**

8. Once a solid has melted, is it possible to change the material back into a solid?

1/3

- **Yes. Freezing the liquid will make the material solid again.**
- No. Melting is a change that can't be reversed.
- Yes, but only if the material is water.

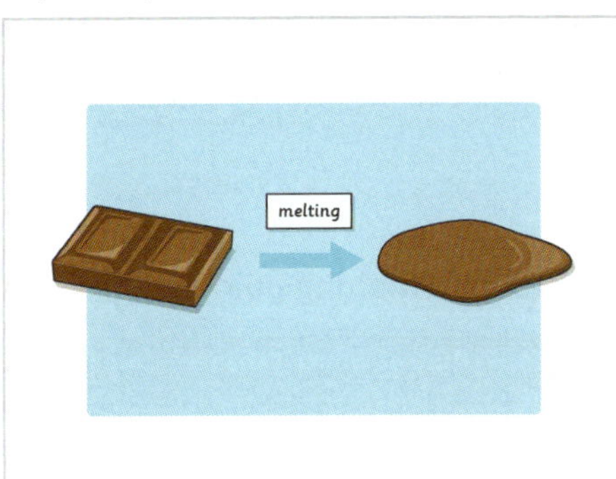

9. Which change of state is sometimes described as solidifying?

1/3

- **freezing**
- melting
- evaporating

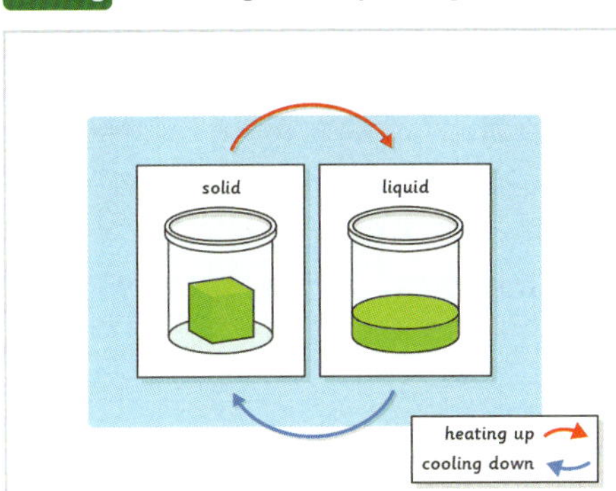

Level 2: Understanding: I can explain that changing temperatures can cause materials to melt or freeze.

Required: 7/10 **Pupil Navigation:** on

Randomised: off

10. The ___________ of a substance is the measure of how hot it is.

- **temperature**
- Celsius
- temperatyur
- temprature
- thermometer
- degrees
- temperatur
- temperachure
- tempratur
- temprachure

11. _______ liquids to their freezing point causes them to freeze.

1/3

- **Cooling**
- Boiling
- Heating

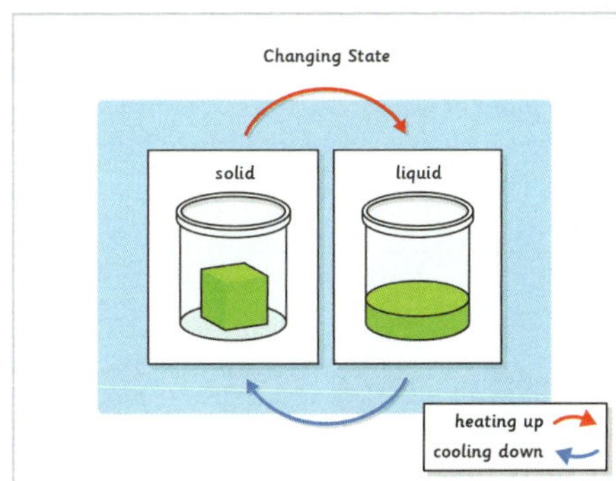

12. You use a ______________ to measure the temperature of something.

1/4

- stop watch
- pipette
- **thermometer**
- ruler

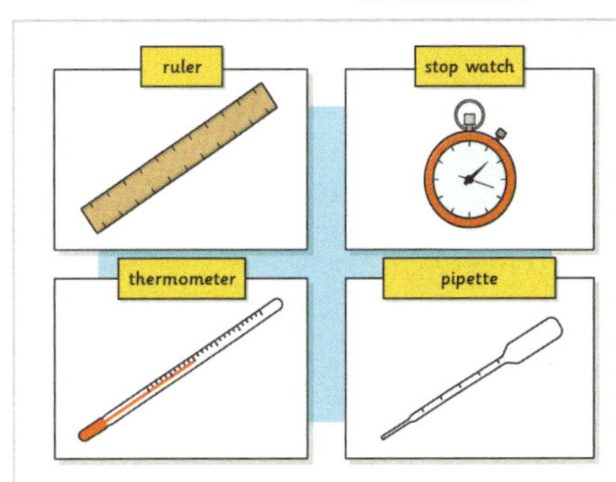

13. Temperature is measured in degrees Celsius. This is usually written as . . .

1/3

- **°C**
- dC
- C

Level 2 *continued*

14. The temperature at which a liquid changes state to become a solid is the substance's . . .

1/3

- boiling point
- freezing point
- turning point

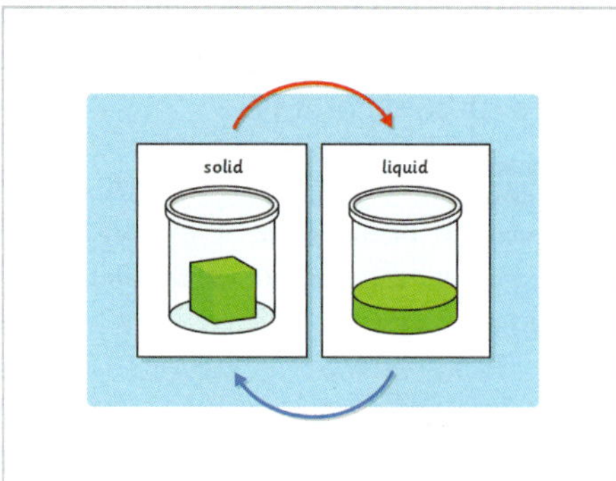

15. 0 °C is the temperature of the . . .

2/3

- melting point of water
- freezing point of water
- boiling point of water

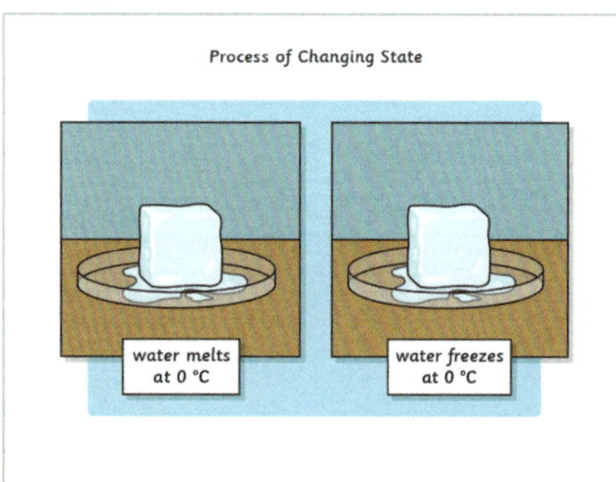

16. The ________ point of a substance is the same temperature as its melting point.

1/3

- boiling
- condensing
- freezing

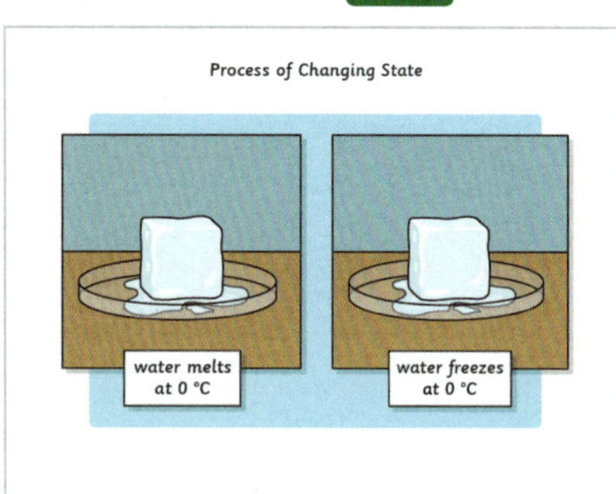

17. Heating a _______ to its melting point causes it to melt.

1/3

- solid
- liquid
- gas

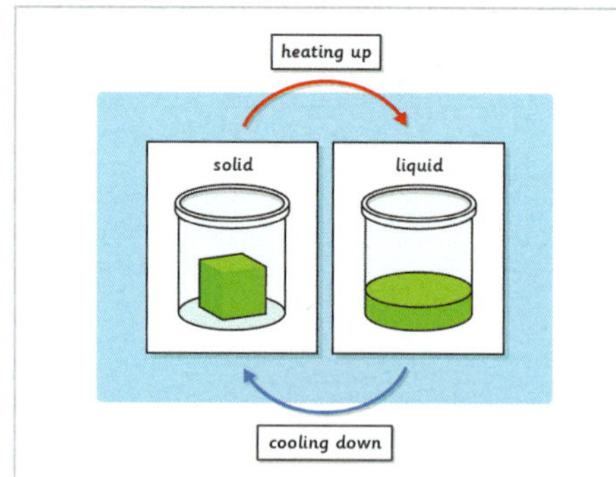

18. What temperature does water freeze at?

1/5

- -18 °C
- 0 °C
- 21 °C
- 37 °C
- 100 °C

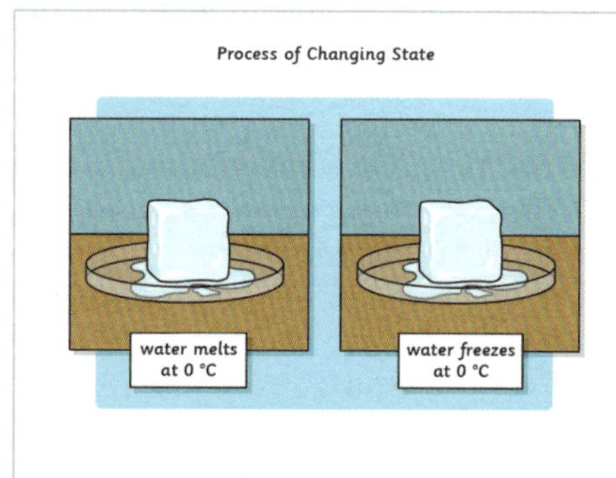

19. A substance's melting point is . . .

1/3

- the temperature at which it changes state from a solid to a liquid.
- the temperature at which it changes state from a liquid to a gas.
- The temperature at which it changes from a gas to a liquid.

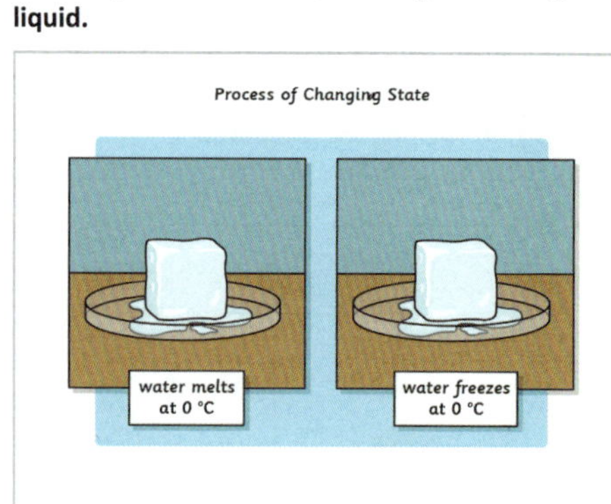

Level 3: Application: I can apply my knowledge and understanding of melting and freezing to the world around me.

Required: 4/4 **Pupil Navigation:** on
Randomised: off

20. Sugar is a liquid at room temperature (21 °C). True or false?

1/2

- true
- **false**

material	melting point temperature (°C)
glass	1400
sugar	185
gold	1064
olive oil	-20
chocolate	35
aluminium	660
salt	800
ice	0
mercury	-39

21. The table shows the temperatures of melting points for different materials. Use the table to select which of these materials are in a liquid state at room temperature (21 °C).

3/7

- glass
- gold
- **olive oil**
- chocolate
- aluminium
- **ice**
- **mercury**

material	melting point temperature (°C)
glass	1400
sugar	185
gold	1064
olive oil	-20
chocolate	35
aluminium	660
salt	800
ice	0
mercury	-39

22. A healthy human has a body temperature of about 37 °C. Use the melting points table to explain why chocolate starts to melt in your hand.

- Open question, no set answer

material	melting point temperature (°C)
glass	1400
sugar	185
gold	1064
olive oil	-20
chocolate	35
aluminium	660
salt	800
ice	0
mercury	-39

23. Why would aluminium not be a good material to use for equipment that is used to melt salt?

- Open question, no set answer

material	melting point temperature (°C)
glass	1400
sugar	185
gold	1064
olive oil	-20
chocolate	35
aluminium	660
salt	800
ice	0
mercury	-39

Processes of Changing State

Objective: I know that some materials change state when they are heated or cooled.

Quick Search Ref: 10638

Level 1: Knowledge: I can describe how some materials can change state.

Required: 5/9 | Pupil Navigation: on | Randomised: off

1. Solid, liquid and gas are all ______ of matter.

a b c

- **states**
- stayts
- state
- staytes

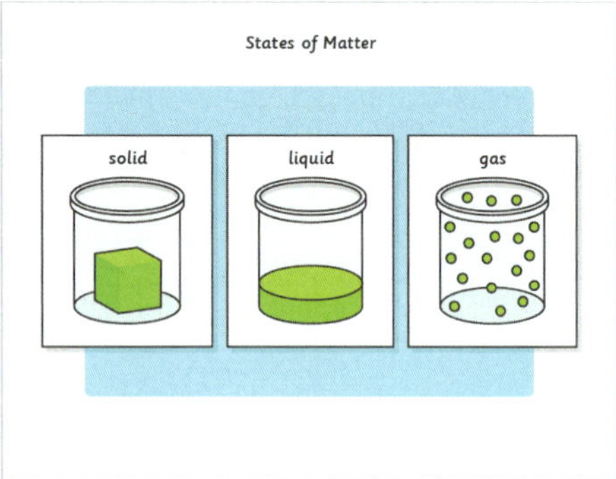

2. Liquids _________ to become gases.

a b c

- melt
- condense
- **boil**
- **evaporate**
- freeze
- evaporating
- evaporation

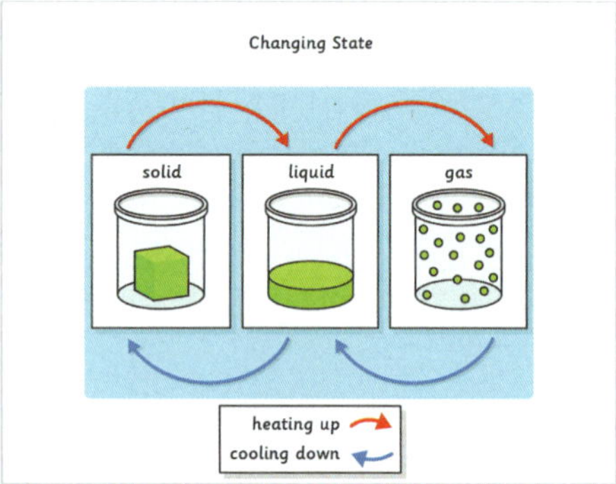

3. What is the opposite process to freezing?

1/3

- evaporating
- **melting**
- condensing

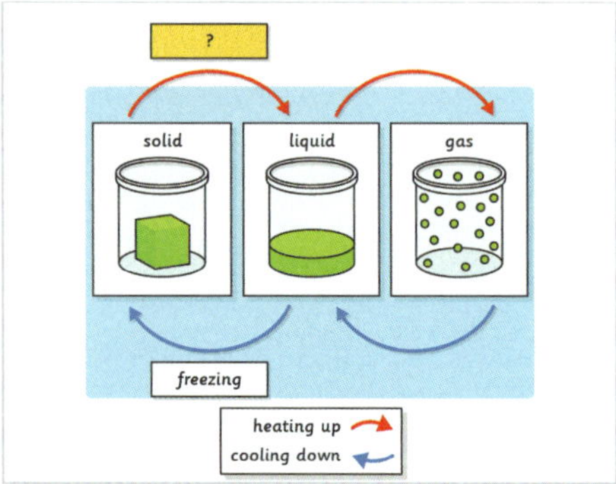

4. A gas that is changing state to become a liquid is . . .

1/4

- melting.
- evaporating.
- **condensing.**
- freezing.

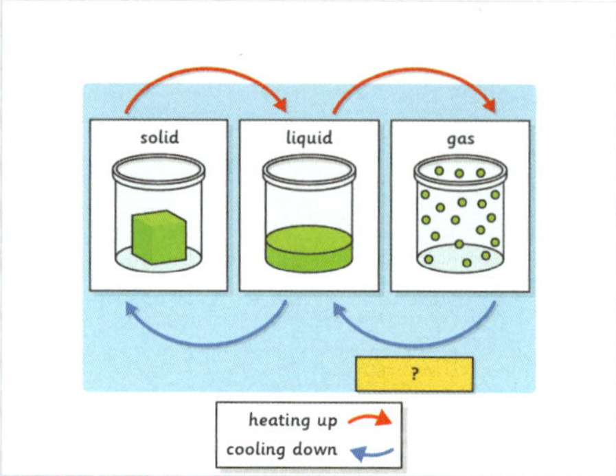

5. ________ is the process of a liquid becoming a solid.

a b c

- frezing
- freesing
- **freezing**
- melting
- condensing
- freazing
- evaporating

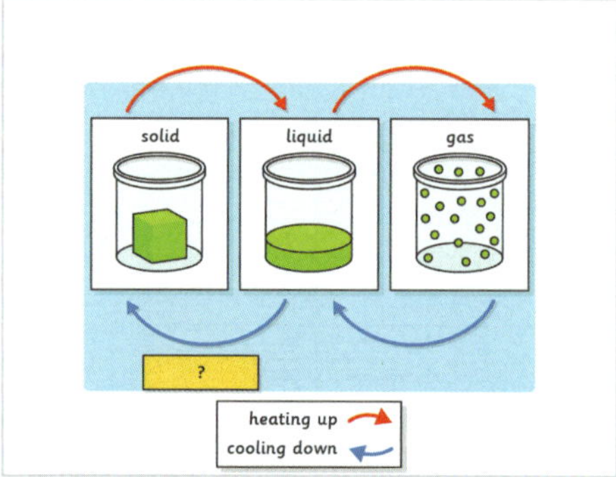

6. Which of the statements about changing states isn't true?

1/4

- **A new material is formed when a substance changes state.**
- Changes of state are reversible.
- Ice and water are the same material in different states.
- When water evaporates, it becomes a colourless, odourless gas.

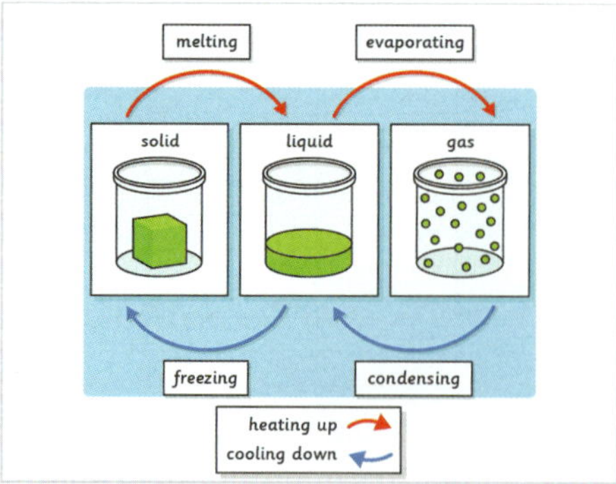

Level 1 *continued*

7. Melting is the change of state from . . .

1/4

- liquid to solid
- gas to liquid
- solid to liquid
- liquid to gas

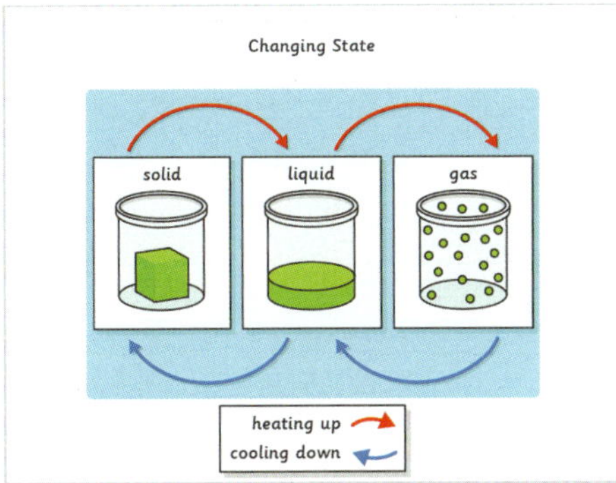

8. What is the opposite process to evaporation?

1/3

- freezing
- melting
- condensation

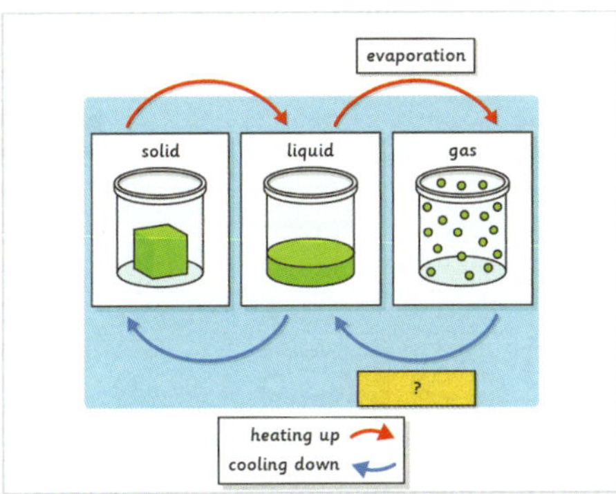

9. Which **two changes in state** result in a liquid?

2/4

- condensation
- freezing
- evaporation
- melting

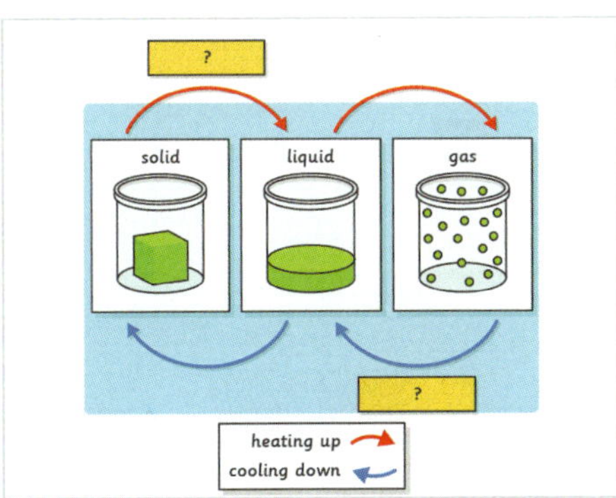

Level 2: Understanding: I can explain that changing temperatures can cause changes of state.

Required: 7/10 **Pupil Navigation:** on

Randomised: off

10. ________ a liquid causes it to evaporate more quickly.

1/3

- Cooling
- Freezing
- Heating

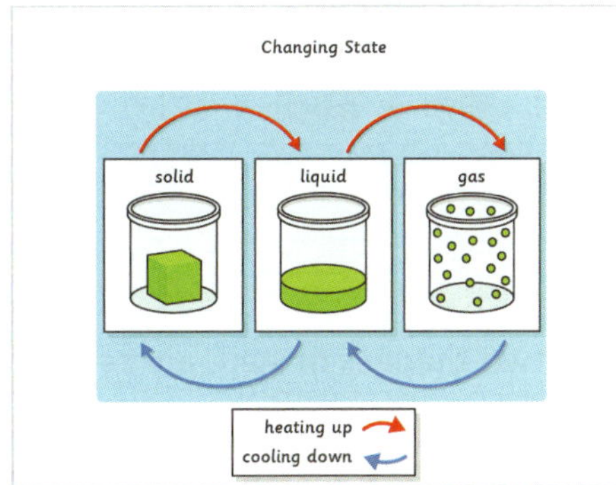

11. What is the temperature of the boiling point of water?

1/5

- -18 °C
- 0 °C
- 21 °C
- 37 °C
- 100 °C

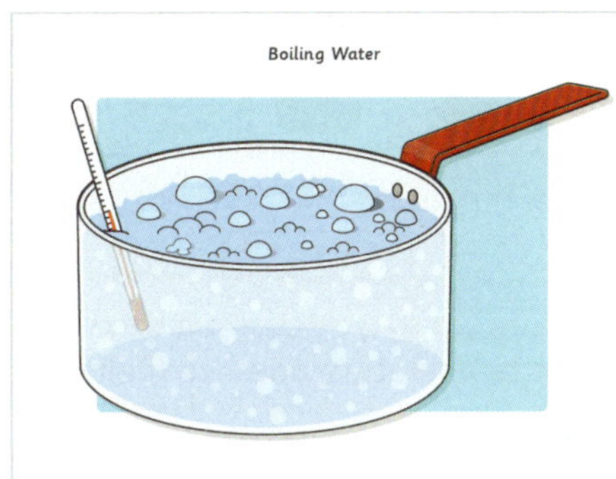

12. ____________ is the change of state caused by cooling gases to their boiling point.

a b c

- Condensation
- Condensing
- Condense

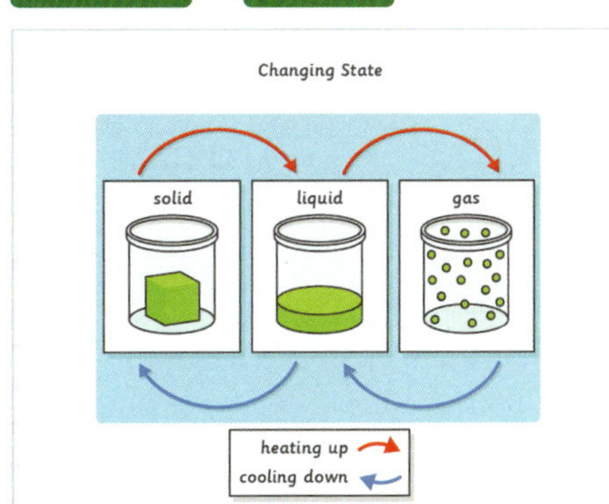

Level 2 *continued*

13. The _ _ _ _ _ _ _ _ point of water is 100 °C

a b c

- boil
- boiling
- boyling
- melting
- evaporating

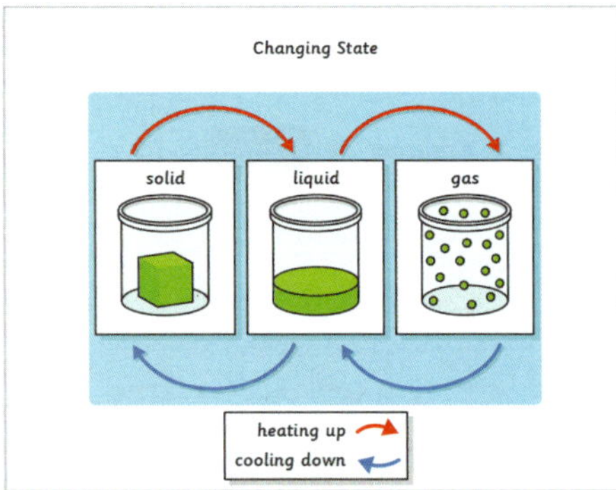

14. Which **two changes of state** are caused by cooling?

2/4

- melting
- freezing
- evaporating
- condensing

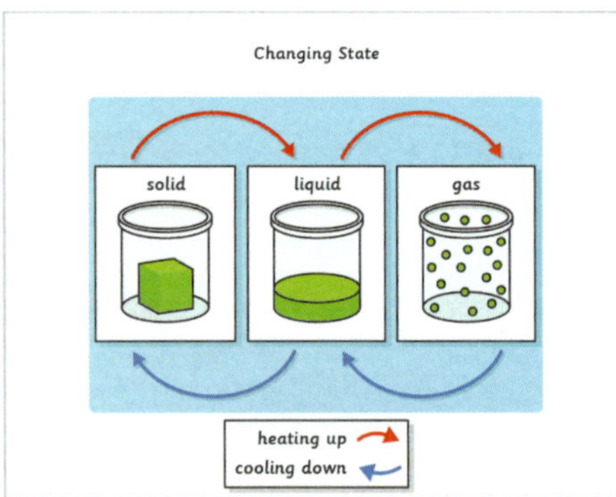

15. At what temperature does ice melt?

1/5

- -18 °C
- 0 °C
- 21 °C
- 37 °C
- 100 °C

16. The _ _ _ _ _ _ _ _ point of a substance is the same temperature as its melting point.

1/3

- boiling
- condensing
- freezing

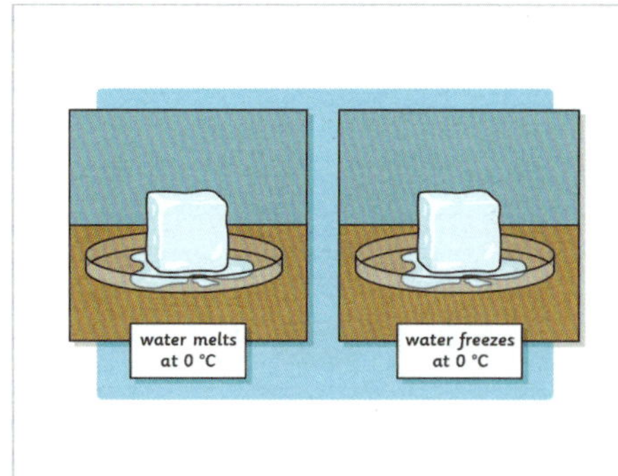

17. Cooling a _______ to its freezing point causes it to freeze.

1/3

- solid
- liquid
- gas

18. Which **two changes of state** are caused by heating?

2/4

- melting
- freezing
- condensing
- evaporating

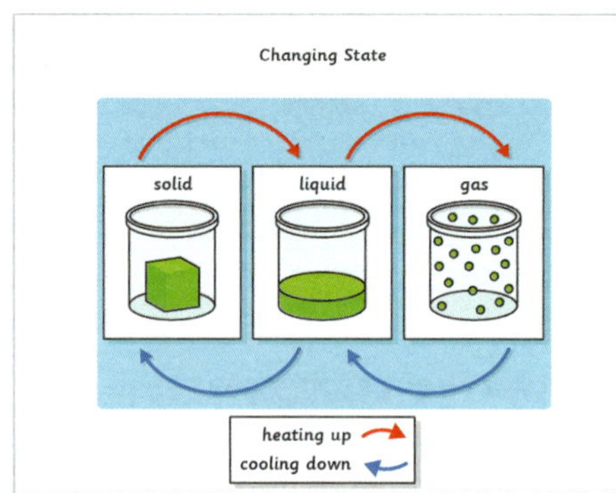

19. What is the temperature of the freezing point of water?

1/5

- -18 °C
- 0 °C
- 21 °C
- 37 °C
- 100 °C

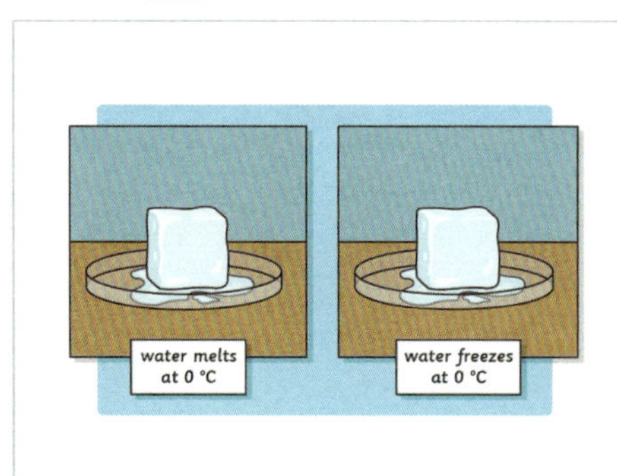

Level 3: Application: I can apply my knowledge and understanding of changing states to the world around me.

Required: 1/3 **Pupil Navigation:** on
Randomised: off

20. Emma, Ryan and Josie have melted different types of chocolate in metal cups floating in a tank of hot water. Who do you think has the best understanding of melting?

1/3

- Emma
- Ryan
- **Josie**

21. Which **two** of these statements about evaporation are incorrect?

2/3

- **When water evaporates, it becomes air.**
- Water starts evaporating before it reaches 100 °C.
- **All liquids evaporate at 100 °C**

22. Kevin has labelled a diagram to show how a cloud is formed. What is wrong with his diagram?

1/4

- Kevin has labelled water vapour as a gas.
- **Kevin has labelled the cloud as being made of water vapour.**
- Kevin has shown water evaporating to form water vapour.
- Kevin has shown the puddle containing liquid water.

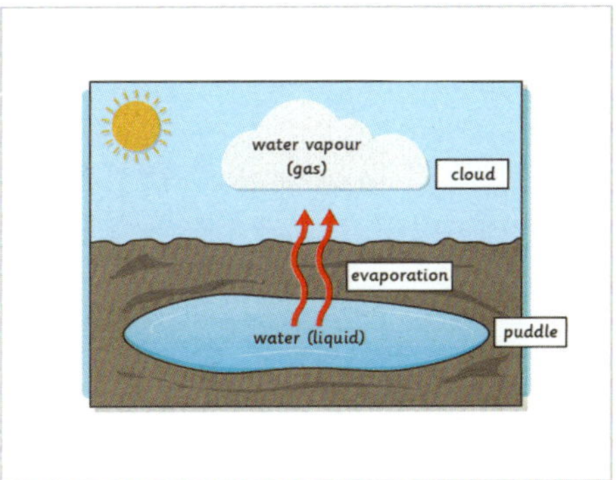

Properties of Solids, Liquids and Gases

Objective: I can compare materials and group them together according to whether they are solids, liquids or gases.

Quick Search Ref: 10642

Level 1: Knowledge: I can describe the properties of solids, liquids and gases.

Required: 6/9 **Pupil Navigation:** on **Randomised:** off

1. Solid, liquid and gas are all ______ of matter.

- stayts
- **states**
- staytes
- state

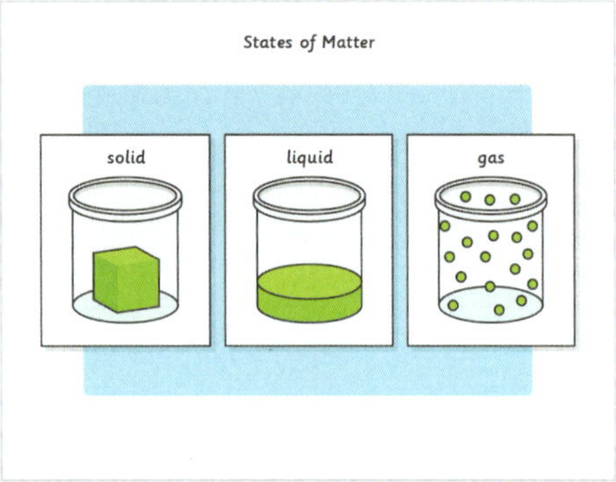

2. Ice is hard, cold and transparent. These are some of the __________ of ice.

1/4

- types
- volume
- states
- **properties**

3. What is the solid state of water called?

1/3

- water
- **ice**
- water vapour

4. Select the properties that describe a solid.

2/5

- **fixed shape**
- flows from one place to another
- **fixed volume**
- takes the shape of its container
- fills the available space in the container

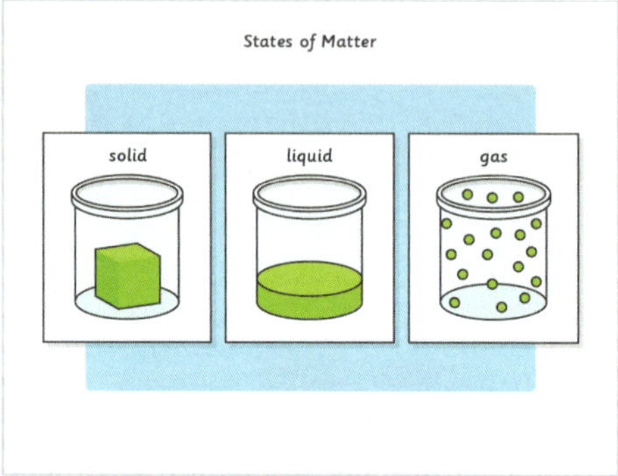

5. Which of these are properties of a liquid?

2/5

- fixed shape
- **flows from one place to another**
- **has a fixed volume**
- **takes the shape of its container**
- fills the available space in the container

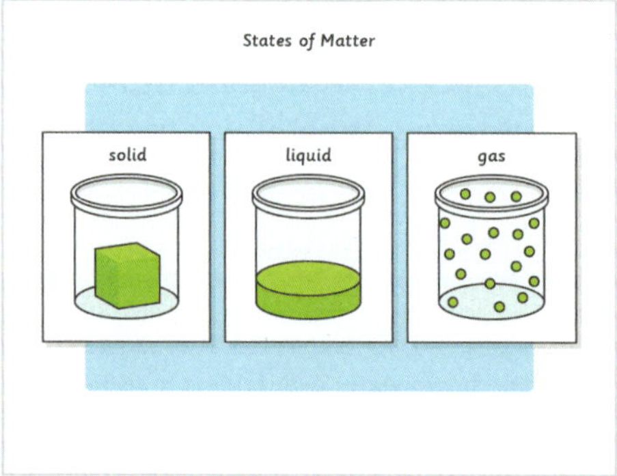

6. A material that flows could be in which states?

1/3

- solid or liquid
- **liquid or gas**
- gas or solid

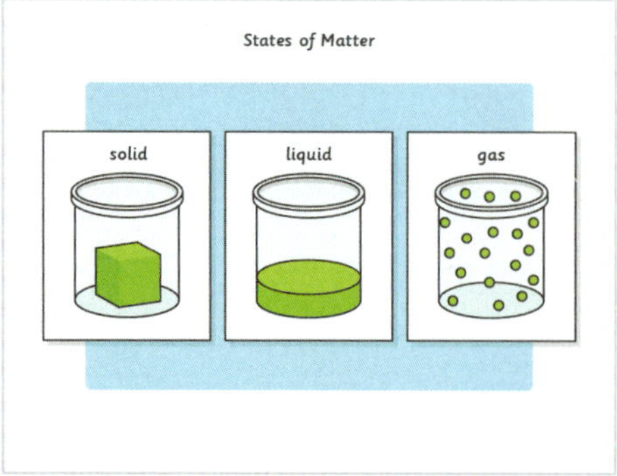

Level 1 *continued*

7. Which of these is a property of a gas, but not a property of a solid or liquid?

1/5

- has a fixed shape
- flows from one place to another
- has a fixed volume
- takes the shape of its container
- **fills the available space in the container**

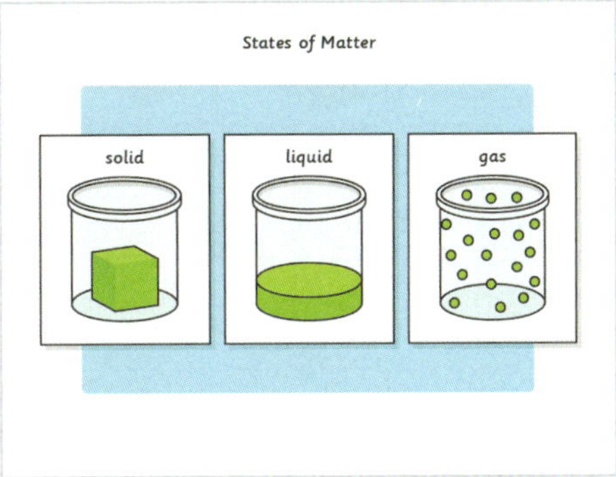

8. A material that doesn't have a fixed volume is in which state?

1/3

- solid
- liquid
- **gas**

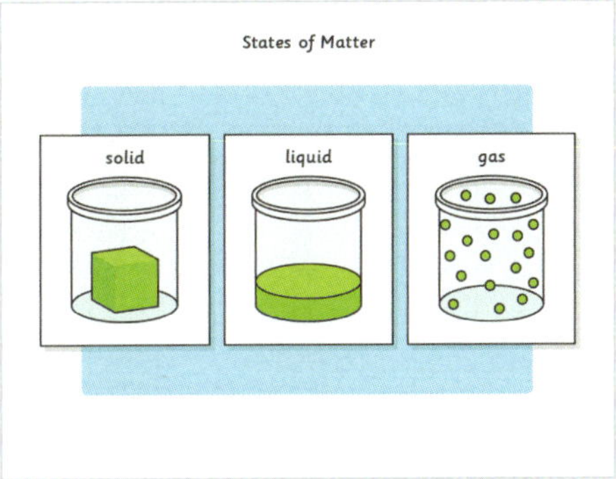

9. A material that has a fixed shape is in which state?

1/3

- **solid**
- liquid
- gas

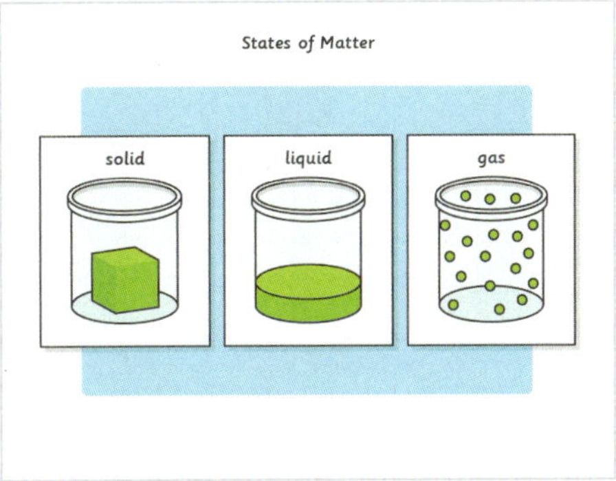

Level 2: Understanding: I can group materials based on their state of matter (solid, liquid, gas).

Required: 6/9 **Pupil Navigation:** on
Randomised: off

10. What are the properties of cooking oil?

3/5

- has a fixed shape
- **flows from one place to another**
- **has no fixed shape**
- fills the available space in the container
- **has a fixed volume**

11. Arrange the states of matter in the order they're shown on the diagram.

- **solid**
- **gas**
- **liquid**

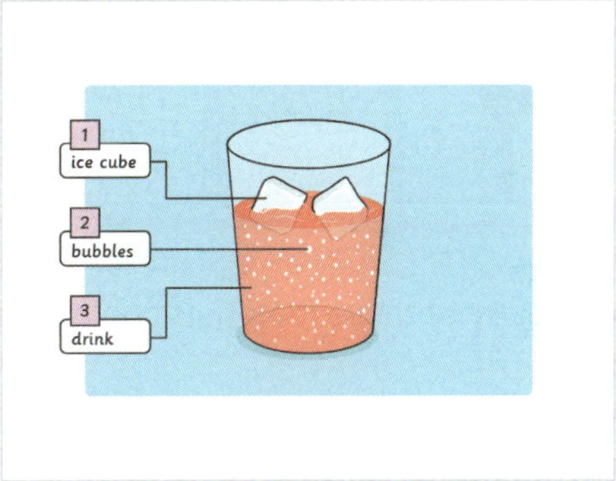

12. Select the properties of a diamond.

2/5

- **has a fixed shape**
- flows from one place to another
- **has a fixed volume**
- takes the shape of its container
- fills the available space in the container

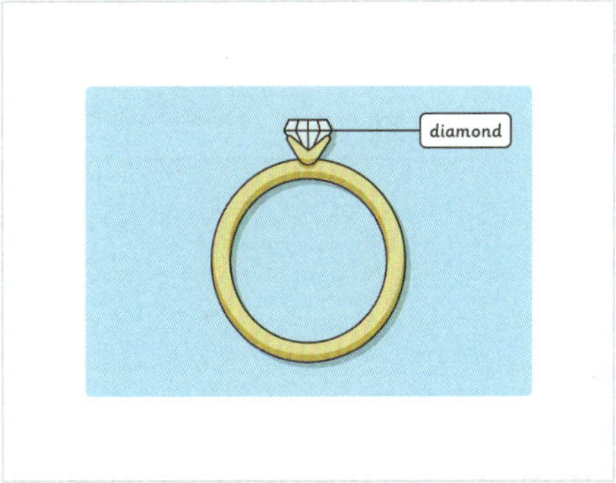

Level 2 *continued*

13. What are the properties of helium?

3/5

- has a fixed shape
- **flows from one place to another**
- has a fixed volume
- **has no fixed shape**
- **fills the available space in the container**

14. Which of these isn't in a solid state?

2/6

- sand
- book
- **ketchup**
- **honey**
- sugar
- cotton wool

15. How many of these materials are in a liquid state?

- **3**
- 4

16. Which of these materials are solid at room temperature?

2/5

- water
- **sponge**
- **gold**
- carbon dioxide
- petrol

17. Which of these materials are liquid at room temperature?

3/5

- **water**
- rock
- air
- **olive oil**
- **shampoo**

18. Which of these materials are gases at room temperature?

2/5

- water
- candy floss
- **air**
- **carbon dioxide**
- sugar

Level 3: Application: I can apply my knowledge and understanding of the states of matter to the world around me.

Required: 3/4 **Pupil Navigation:** on
Randomised: off

19. Kevin poured one litre of coloured liquid from a cylindrical container into a cuboid container. He recorded his investigation in pictures. Which picture shows what the liquid looked like in the cuboid container?

1/3

- a)
- b)
- **c)**

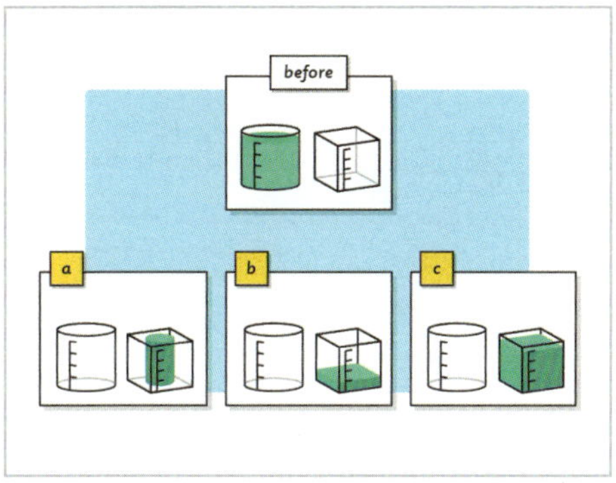

20. Steve fills a plastic syringe with 20 ml of air. He puts his finger over the open end and pushes down the plunger. When Steve can't push the plunger any further, the scale shows there is 5 ml of air in the syringe. Why is the volume of air smaller at the end of the investigation?

1/3

- The air escaped from the end of the syringe.
- **The air in the syringe was squashed, so it had a smaller volume.**
- The air leaked out of the syringe around the plunger.

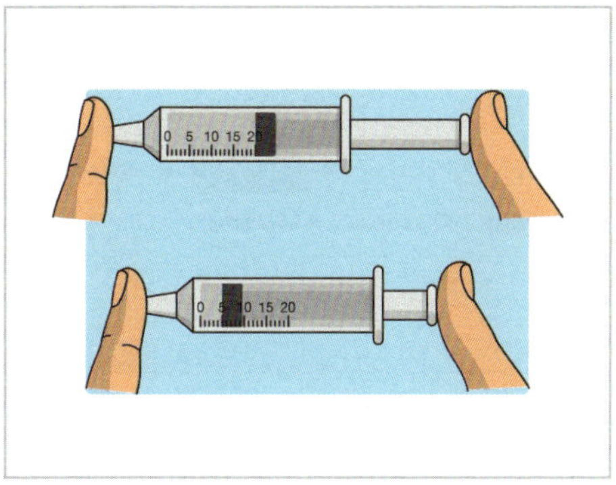

21. How many different materials are shown in the pictures?

- **1**
- 5

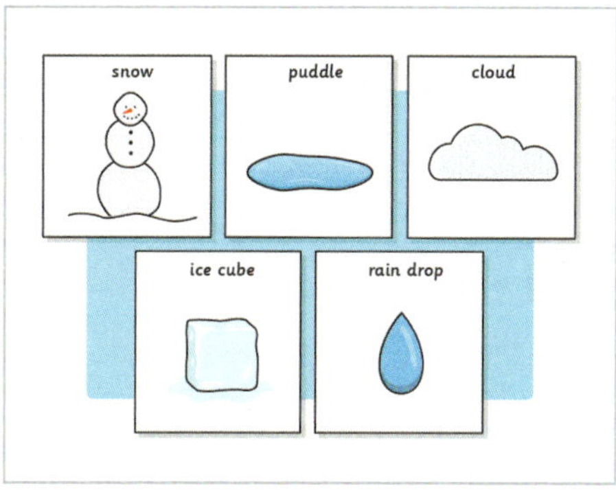

22. The solid chocolate has turned into a liquid in Penny's hand. What has happened to the chocolate?

1/3

- **It has changed state.**
- It has changed into a different type of chocolate.
- It has been squashed.

States of Matter Vocabulary

Objective: I can use the key words associated with states of matter.

Quick Search Ref: 10629

Level 1: Knowledge: I can spell and define scientific words for states of matter.

Required: 13/13 **Pupil Navigation:** on **Randomised:** off

1. Solid, liquid and gas are all _ _ _ _ _ _ of matter.

abc

- **states**
- staytes
- state
- stayts

2. Sticky, runny, shiny and translucent are some of the __________ of honey.

1/3

- states
- types
- **properties**

3. Ice is water in its _ _ _ _ _ state.

abc

- **solid**

4. A solid becomes a _ _ _ _ _ _ when it melts.

abc

- likwid
- **liquid**
- licwid
- liqid
- water
- lickwid

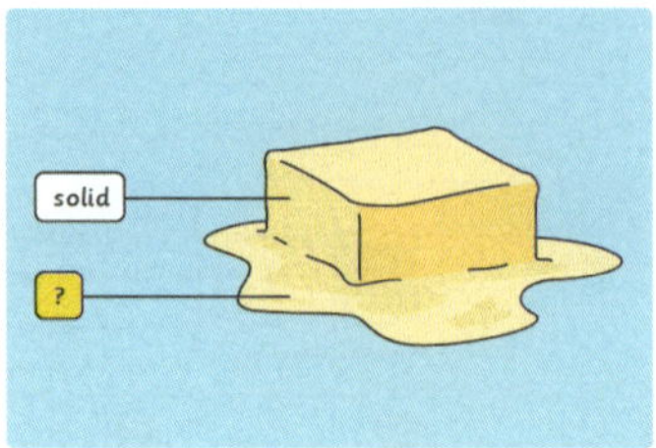

5. When water evaporates, it becomes a _ _ _ called water vapour.

abc

- **gas**
- gass
- gaz

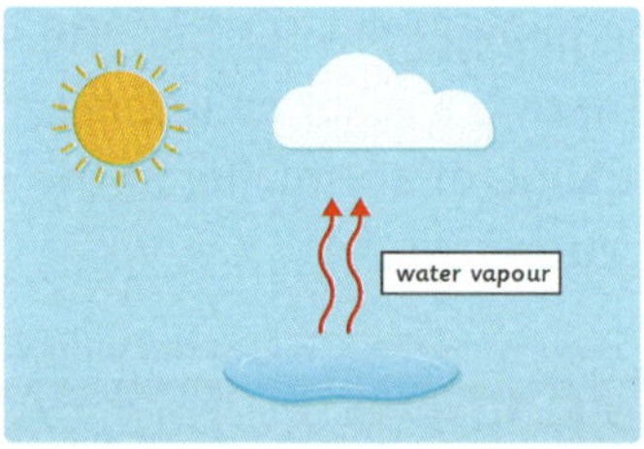

6. _ _ _ _ _ _ _ _ _ is the process of changing state from a liquid to a solid.

abc

- Freezes
- Freazing
- **Freezing**
- Freesing
- Solidifying
- Freeze
- Frezing

Level 1 *continued*

7. A solid _ _ _ _ _ to become a liquid.

abc

- **melts** ■ **melt**

8. _ _ _ _ _ _ _ _ _ _ _ _ is the change of state from liquid to gas.

abc

- **evaporation** ■ **evaporates** ■ **evapourasion**
- **evaporasion** ■ **evaporashun** ■ **evaporate**
- **evapourashun** ■ **evapouration**

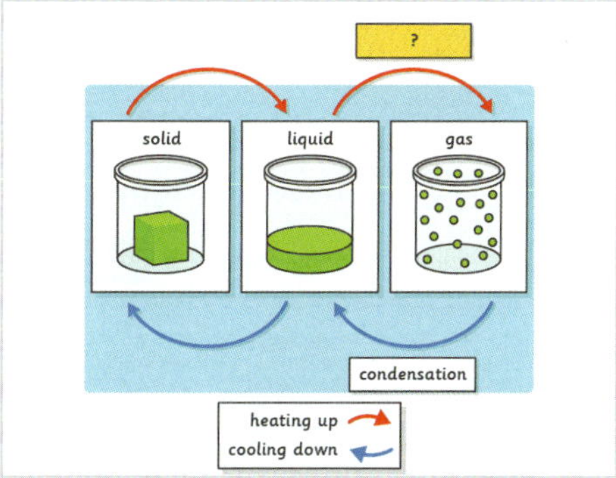

9. _ _ _ _ _ _ _ _ _ _ _ _ is the opposite process to evaporation.

abc

- **condensation** ■ **condensate** ■ **condense**
- **condensing** ■ **condenses** ■ **condensasion**
- **condensashun**

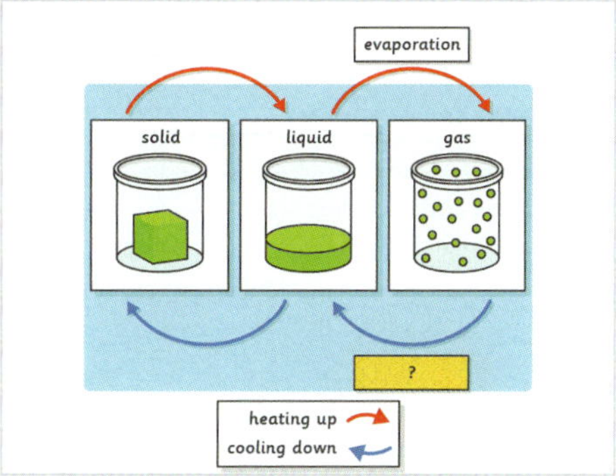

10. Water on Earth evaporates into water vapour, condenses to form clouds and then falls back to the ground as rain or snow. This process is known as . . .

☐ ☒ ☐ 1/3

- **the water cycle** ■ **weather** ■ **precipitation**

11. The _ _ _ _ _ _ _ _ _ _ _ of a cup of coffee is the measurent of how hot the coffee is.

abc

- **Celsius** ■ **temperature** ■ **degrees** ■ **temprachure**
- **temprature** ■ **degrees Celsius** ■ **thermometer**
- **temperachure**

12. The equipment used to measure temperature is a . . .

☐ ☒ ☐ 1/4

- **ruler** ■ **stop watch** ■ **thermometer**
- **measuring cylinder**

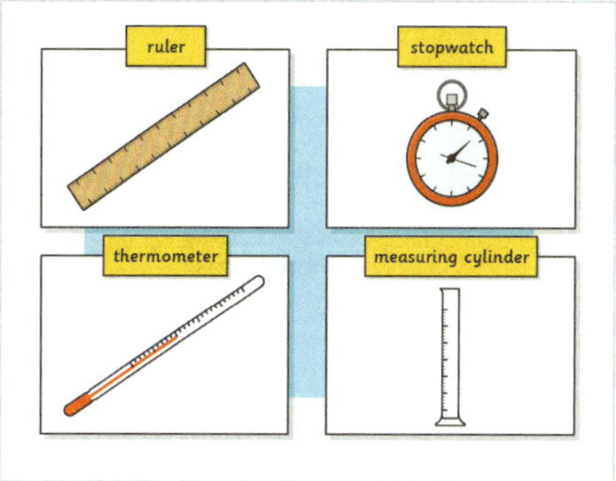

13. The _ _ _ _ _ _ of a solid, liquid or gas is the amount of space it takes up.

abc

- **volume** ■ **millilitres** ■ **cubic centimetres** ■ **litres**
- **centimetres cubed** ■ **centimetres**

The Water Cycle

Objective: I can understand the role of evaporation and condensation in the water cycle.

Quick Search Ref: 11250

Level 1: Knowledge: I can describe water in three different states of matter and the changes in state that water undergoes.

Required: 6/8 | Pupil Navigation: on | Randomised: off

1. Matter is . . .

1/3

- whether something is a solid, liquid or gas.
- **the 'stuff' that everything is made up of.**
- how much space something takes up.

2. Solid, liquid and gas are all _ _ _ _ _ _ of matter.

- staytes
- **states**
- state

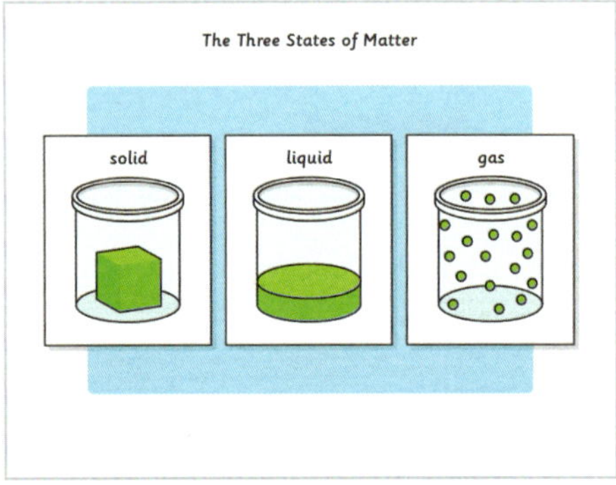

3. When ice melts, it becomes liquid _ _ _ _ _ .

- **water**
- worter
- watr
- warter

4. When water evaporates, it becomes . . .

1/4

- air.
- oxygen.
- **water vapour.**
- helium.

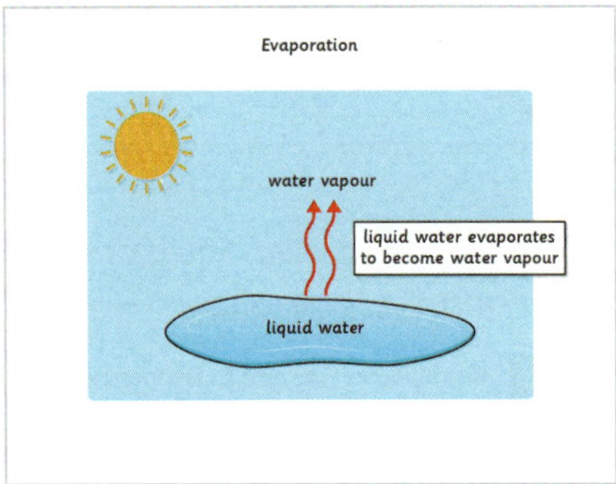

5. Water vapour is a __________ and _______ gas.

2/5

- **colourless**
- smelly
- smoky
- **odourless**
- blue

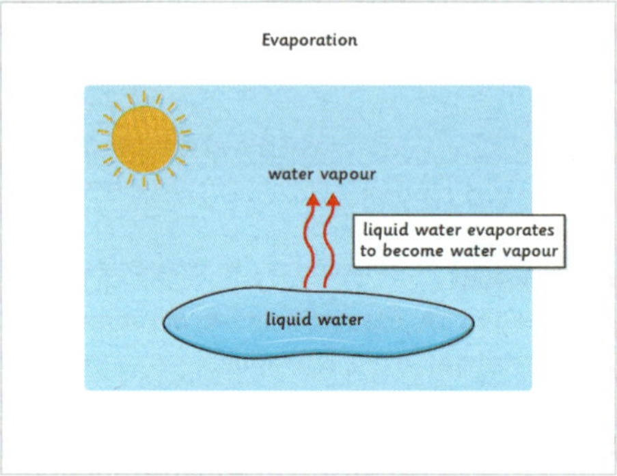

6. When water vapour is cooled, it ___________ and becomes liquid water.

1/4

- freezes
- melts
- evaporates
- **condenses**

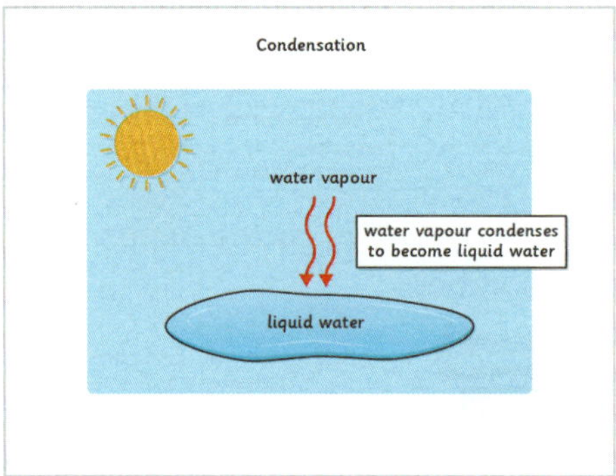

7. As a puddle dries, the water . . .

1/3

- evaporates and disappears.
- **evaporates and becomes water vapour.**
- evaporates and becomes air.

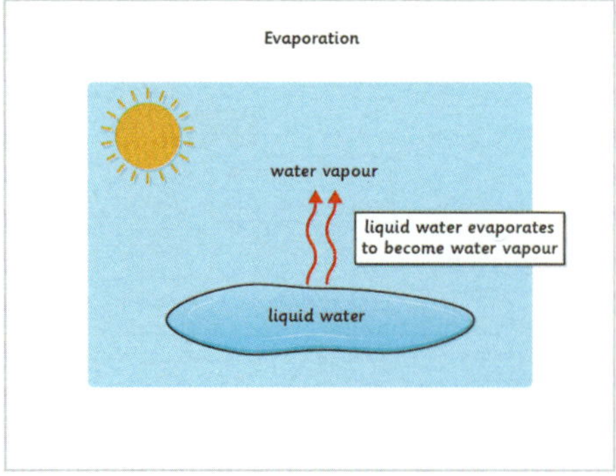

Level 1 *continued*

8. ___ is solid water.

abc

- **ice**
- **ise**

Level 2: Understanding: I can explain the parts played by evaporation and condensation in the water cycle.

Required: 7/9 **Pupil Navigation:** on

Randomised: off

9. Arrange the stages of the water cycle into the correct order. The first step has been numbered for you.

↑↓

- **1. Water evaporates from the sea and the water vapour rises.**
- **Water vapour cools and condenses into tiny water droplets, forming clouds.**
- **Droplets of water in clouds join together.**
- **Water in the clouds falls as rain.**
- **Rainfall collects in lakes and rivers.**
- **The water in rivers flows into the sea.**

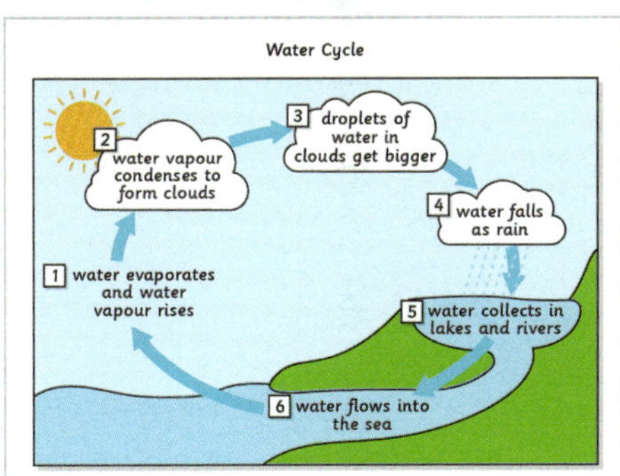

10. Who has the best understanding of what clouds are made of?

1/3

- **Nathan**
- **Leah**
- **Zach**

11. In stage 1 of the diagram, water __________ from the sea.

1/4

- **melts**
- **condenses**
- **evaporates**
- **freezes**

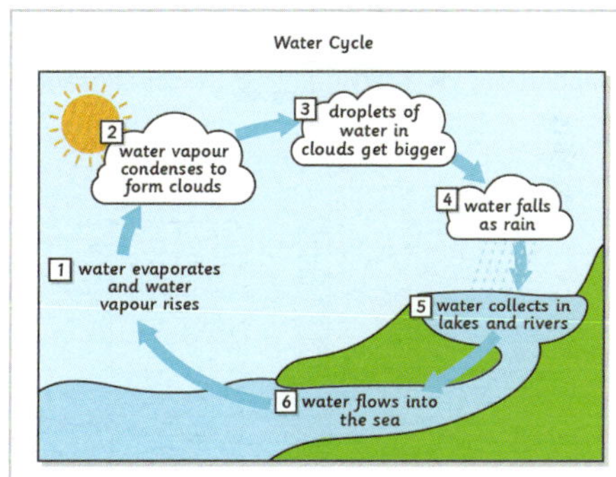

12. In stage 1 of the water cycle diagram, water evaporates from the sea and becomes a gas called ______________.

1/4

- **helium**
- **carbon dioxide**
- **water vapour**
- **air**

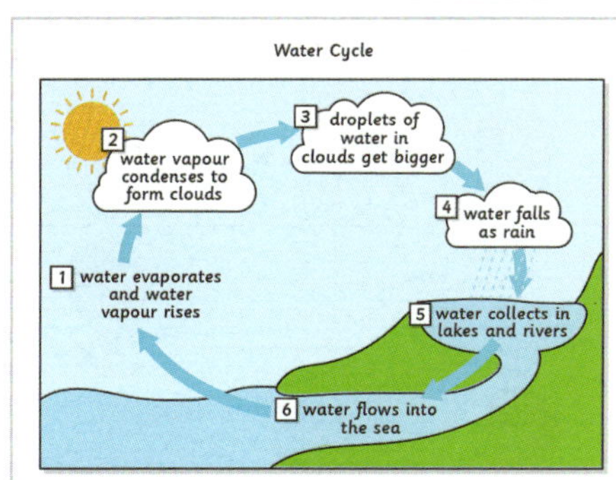

Level 2 *continued*

13. Clouds form when water vapour ________.

1/3

- evaporates
- freezes
- **condenses**

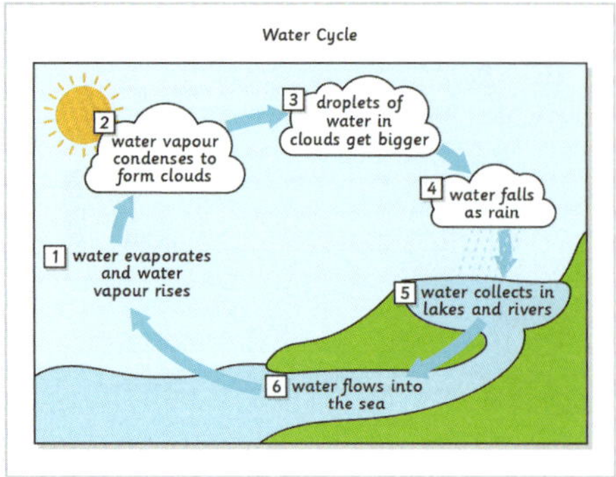

14. Water vapour condenses into droplets of water that fall as rain. When it's cold, the droplets form hail, sleet and snow. These are all types of . . .

1/3

- evaporation.
- **precipitation.**
- condensation.

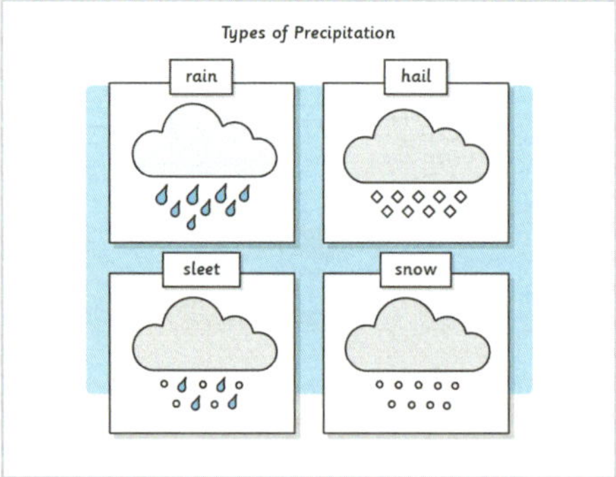

15. Grace, Lewis and Liam are talking about the water cycle. Who has the right idea?

1/3

- Grace
- **Lewis**
- Liam

16. Which of these statements about the water cycle is incorrect?

1/3

- **The water cycle ends when the water flows into the sea.**
- The water on Earth is continually recycled through the water cycle.
- Water evaporates from land, lakes and rivers as well as the sea.

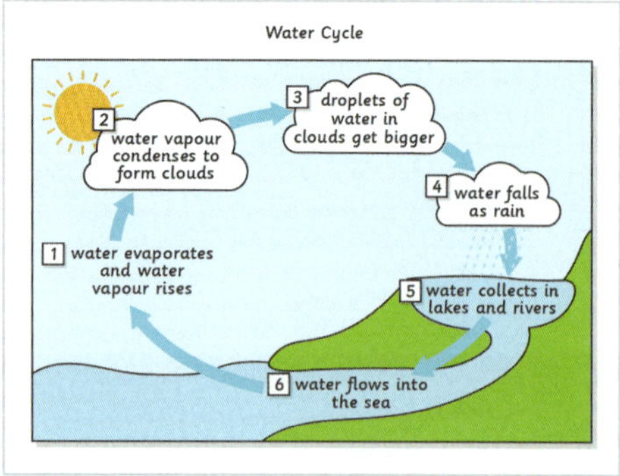

17. In stage 3 of the diagram, water droplets in a cloud join together, making heavier droplets. The water droplets then . . .

1/3

- **fall as rain.**
- evaporate again.
- disappear.

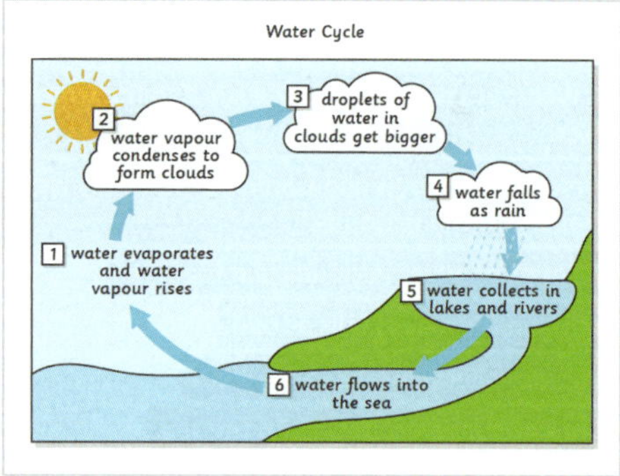

Level 3: Application: I can link the processes of the water cycle to a model.

Required: 1/2 **Pupil Navigation:** on
Randomised: off

18. To make a bottle garden, Eve puts gravel and soil in the bottom of a plastic bottle and plants a plant in the soil. She adds water and seals the bottle with a screw top. After a while, she notices mist forming on the inside of the bottle. What has happened inside Eve's bottle?

1/3

- **The water has evaporated from the soil and the mist is water vapour.**
- **Eve has breathed on the outside of the bottle and her breath has condensed on the bottle.**
- **The water has evaporated from the soil and condensed on the side of the bottle.**

19. Eve makes a sealed bottle garden. She has written a description of what is happening inside the bottle, but it is in the wrong order. Arrange Eve's labels from 1 to 5 to match the numbers on the diagram.

- **Water in the soil evaporates.**
- **Warm air containing water vapour rises up inside the bottle.**
- **Water vapour cools and condenses to form a mist on the side of inside of the bottle.**
- **Tiny droplets of water join together and get heavier.**
- **The heavier droplets of water run down the bottle and back into the soil.**

Primary Science Y3/4

Earth

Fossils
Properties of Rock

Fossil Formation

Objective: I can describe in simple terms how fossils are formed.

Quick Search Ref: 10669

Level 1: Knowledge: I can identify fossils and recall facts about them.

Required: 7/10 **Pupil Navigation:** on **Randomised:** off

1. What are fossils? 1/3

- dinosaur bones that are over 10,000 years old
- **remains of plants or animals that are over 10,000 years old**
- dead animals that were turned to stone over 10,000 years ago

2. Which of these things can form fossils? Select **one** answer. 1/5

- none of these things
- flowers
- fish
- eggs
- **all of these things**

3. Most plant and animal remains don't form fossils because they _ _ _ away or are eaten before they can fossilise.

- **rot**
- die

4. Jax, Anna and Seth are talking about fossils. Who is correct? 1/3

- Jax
- **Anna**
- Seth

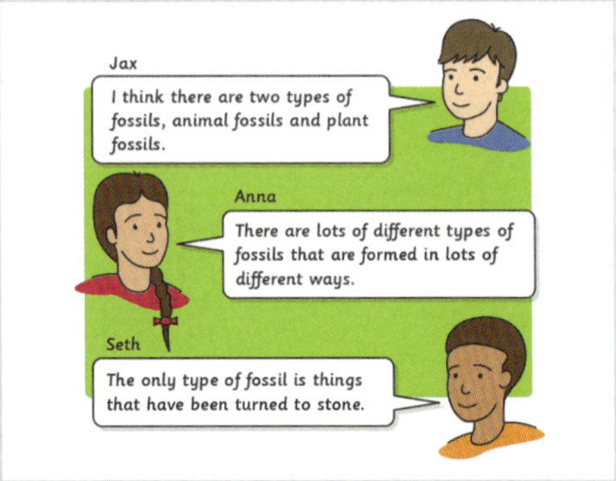

5. Arrange the animal fossils in the order that they are numbered in the image.

- **fish fossil**
- **snail fossil**
- **dinosaur fossil**
- **insect fossil**

dinosaur by Matthias Kabel; *fish* by Michael Popp; *snail*, *insect* by James St. John

6. A scientist who studies fossils is called . . . 1/3

- a biologist.
- **a paleontologist.**
- an archaeologist.

7. How old do remains of animals and plants have to be before they are called fossils? 1/3

- 100 years old
- **10,000 years old**
- 1,000,000 years old

8. Fossils are . . . 1/3

- dead animals that have been trapped in ice for over 10,000 years.
- animal bones that are over 10,000 years old.
- **the remains of plants or animals that are over 10,000 years old.**

9. Jess is looking at the shell of an animal that is about 800 years old. Is it a fossil? 1/2

- Yes, it is a fossil.
- **No, it isn't a fossil.**

10. Which of these things can form fossils? Select **one** answer. 1/6

- none of these things
- insects
- leaves
- branches
- reptiles
- **all of these things**

Level 2: Understanding: I can describe how fossils are formed.

Required: 6/8 **Pupil Navigation:** on **Randomised:** off

11. Fossils can't be formed everywhere. Where are most mould and cast fossils formed? 1/3

- in a forest
- **under the sea**
- in the desert

trilobite cast and mold fossil by James St. John

Level 2 *continued*

12. What type of rock are mould and cast fossils formed in?

1/3

- igneous
- metamorphic
- sedimentary

13. Order the steps to show how a mould and cast fossil is formed in sedimentary rock.

- An animal dies and sinks to the sea floor.
- The animal's skin and flesh rot away.
- The skeleton is buried by layers of sediment.
- The shape of the skeleton forms a mould.
- Minerals and sediment fill the mould, forming a cast.
- The fossil comes to the surface.

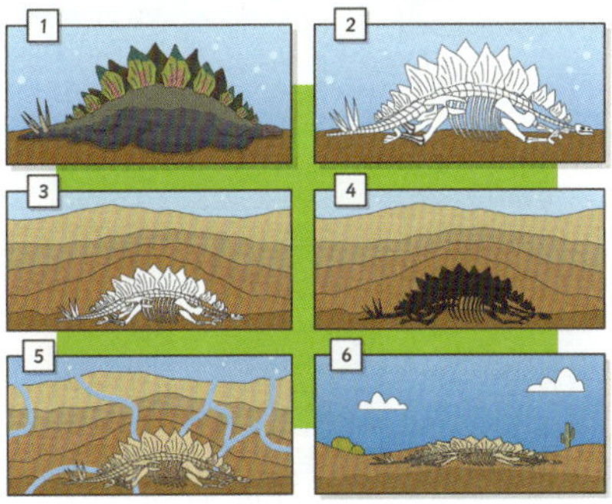

14. Some fossils are preserved in amber. Amber is a sticky liquid from a tree that covers a plant or animal and then turns solid. Which of these fossils are most likely to have been found in amber?

1/3

- insects
- leaves
- sharks

amber by James St. John

15. Sometimes fossils have been completely preserved in ice. Which one of these animal fossils is most likely to have been found preserved in ice?

1/4

- tyrannosaurus rex
- lion
- woolly mammoth
- whale

16. If a plant or animal dies in a hot, dry place it can form a mummified fossil. Where is this most likely to happen?

1/3

- in a forest
- under the sea
- in the desert

sleeping gorgon by James St. John

17. Most fossils that have been discovered are mould and cast fossils. Where are most of these fossils formed?

1/3

- under the sea
- in the desert
- in ice

fossil by Heribert Bechen

18. Mould and cast fossils are formed in __________ rock.

1/3

- metamorphic
- sedimentary
- igneous

Level 3: Working scientifically: I know how to make fossil models.

Required: 4/4 **Pupil Navigation:** on
Randomised: off

19. Frankie is using a raisin to model a type of fossil. She melts a boiled sweet, pours the liquid over the raisin and lets it harden. What type of fossil is she modelling?

1/4

- mould and cast fossil
- trace fossil
- **fossil preserved in amber**
- fossil preserved in ice

20. Frankie is using a strawberry to model a type of fossil. She places the strawberry in container of water and puts it in the freezer. What type of fossil is she modelling?

1/4

- mummified fossil
- trace fossil
- **fossil preserved in ice**
- mould and cast fossil

21. Frankie is using a cherry to model a type of fossil. She has covered the cherry in marzipan and left it to dry. When she removes the marzipan, she can see an indentation of the cherry in it. What type of fossil is she modelling?

1/4

- trace fossil
- mummified fossil
- **mould and cast fossil**
- fossil preserved in ice

22. Frankie is using an apple to model a type of fossil. She places the apple in an oven on a low heat for 10 hours. What type of fossil is she modelling?

1/4

- mould and cast fossil
- trace fossil
- **mummified fossil**
- fossil preserved in amber

Level 4: Extended reasoning: I can apply my knowledge and understanding of fossil formation to the world around me.

Required: 3/3 **Pupil Navigation:** on
Randomised: off

23. Sometimes we find trace fossils of things left behind by dinosaurs. What is this a fossil of?

1/3

- dinosaur footprint
- **dinosaur poo**
- dinosaur tooth

24. Mikey, Becca and Anil are talking about how dinosaurs became extinct. Who is correct?

1/3

- Mikey
- **Becca**
- Anil

25. Mary Anning is famous for discovering lots of dinosaur fossils. Which nursery rhyme was written about Mary Anning?

1/3

- Ring-a-ring O' Roses
- Row, Row, Row Your Boat
- **She Sells Sea Shells**

Mary Anning by Mr. Grey

Rock Properties and Uses

Objective: I can group together different kinds of rocks on the basis of their appearance and simple physical properties.

Quick Search Ref: 10927

Level 1: Knowledge: I can describe the properties of rocks.

Required: 8/10 **Pupil Navigation:** on **Randomised:** off

1. Which of these things has never been alive?

1/3

- sunflower
- rabbit
- **rock**

sunflower by Jon Sullivan

2. Rock is a natural substance that is made of ___________.

1/3

- mud
- sand
- **minerals**

cliff by itravelNZ

3. The properties of a rock describe . . .

1/3

- its name.
- its colour.
- **how it looks or behaves.**

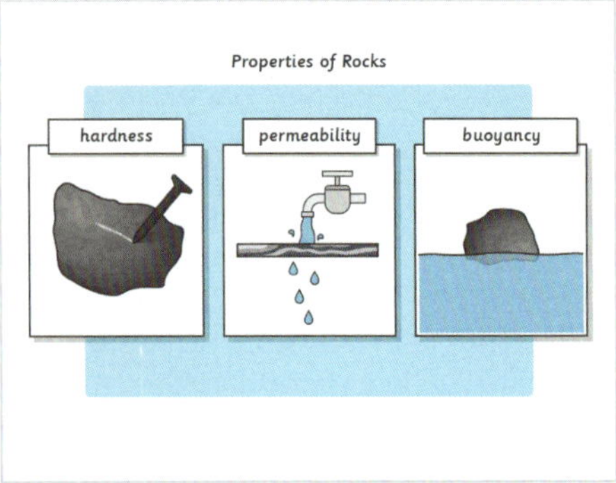

4. Hardness is a property of rocks. The harder the rock, . . .

1/4

- the more difficult it is for water to flow through it.
- the more easily water can flow through it.
- **the more difficult it is to scratch.**
- the easier it is to scratch.

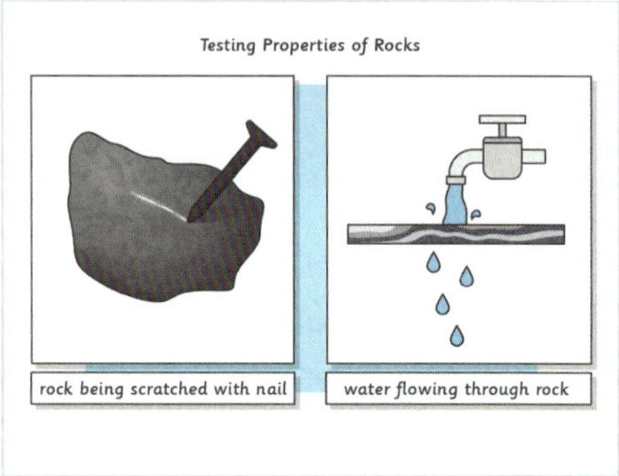

5. Ella, Jude and Amar are discussing whether or not buoyancy is a property of all rocks. Who do you think has the right idea?

1/3

- Ella
- **Jude**
- Amar

Level 1 *continued*

6. Some rocks are permeable, which means that . . .

1/3

- they can't be easily scratched.
- **water can flow through them.** ✓
- they float.

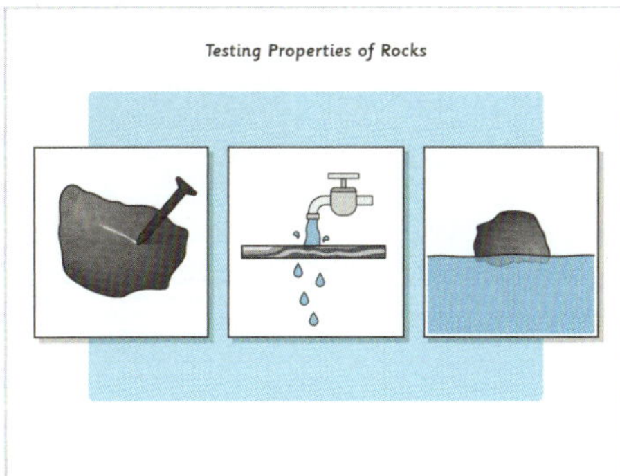

7. Water can't flow through ______________ rocks.

1/3

- permeable
- **impermeable** ✓
- hard

8. Rocks are formed from ________ substances called minerals.

1/3

- **natural** ✓
- metal
- man-made

minerals by Linda Spashett

9. Permeability describes . . .

1/3

- how easy it is to scratch a material.
- **how easy it is for water to flow through a material.** ✓
- how well a material floats.

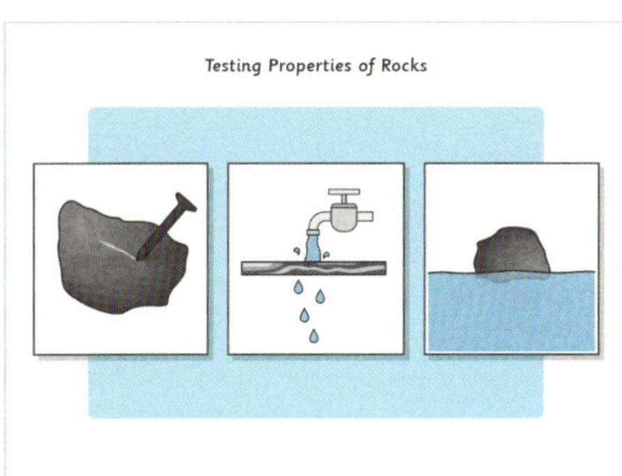

10. The ______ the rock, the more difficult it is to scratch.

a b c

- **hard** ✓
- **harder** ✓
- softer
- hardr

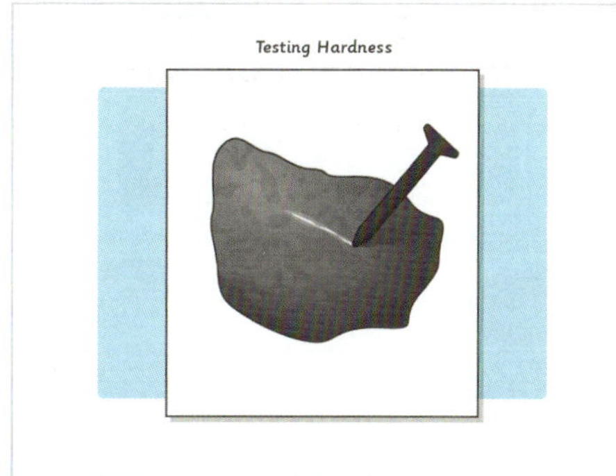

Level 2: Understanding: I can describe the different uses of rocks, giving reasons. I can identify rocks based on their appearance and physical properties.

Required: 6/6 **Pupil Navigation:** on

Randomised: off

11. What are the **three** main types of rock?

3/5

- **sedimentary** ✓
- hard
- clay
- **igneous** ✓
- **metamorphic** ✓

12. Which **two** statements describe sedimentary rocks?

2/4

- **They sometimes contain fossils.** ✓
- They contain lots of holes made by gas bubbles.
- They are formed from molten rock.
- **They are formed in layers.** ✓

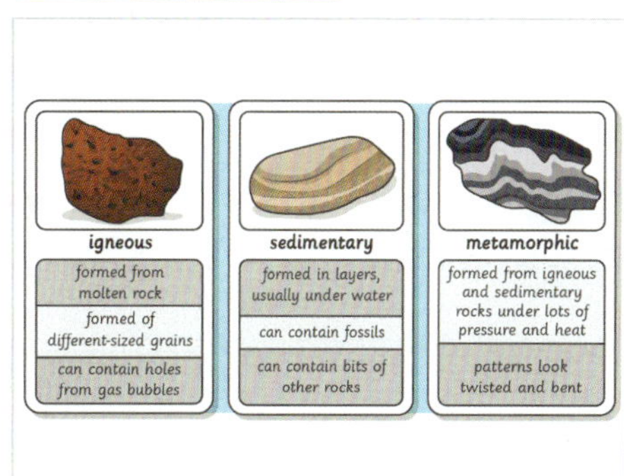

Level 2 *continued*

13. Use the key to identify what type of rock marble is.

1/3

- igneous
- sedimentary
- **metamorphic**

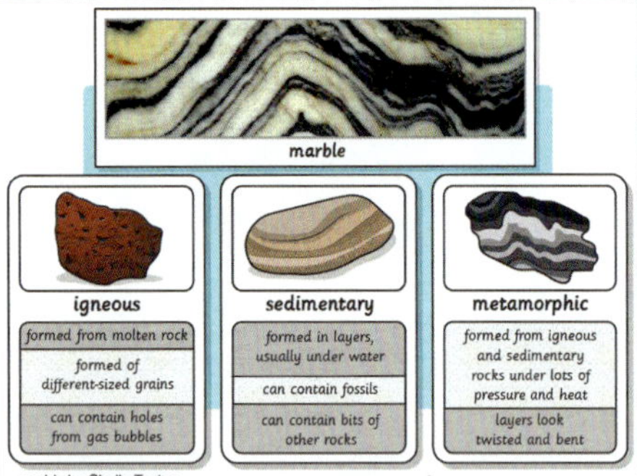

marble by Sheila Tostes

14. Which **two** properties of slate make it a good material for tiling roofs?

2/4

- **impermeability**
- stretchiness
- **strength**
- permeability

slate by Erich Ferdinand; *slate roof* by Bryn Pinzgauer

15. Kitchen worktops need to be hard, strong, impermeable and smooth. Which rock would be best material to make a kitchen worktop from?

1/3

- **granite**
- sandstone
- shale

granite by James Bowe; *sandstone* by Jared Tarbell; *shale* by Kyle Hartshorn

16. You can use chalk to write on different surfaces. Which property makes chalk a good material for writing?

1/3

- permeability
- smoothness
- **softness**

chalk by Amanda Slater; *drawing chalk* by Steven Depolo

Level 3: Application: I can apply my knowledge and understanding of rock properties and uses to the world around me.

Required: 2/3 **Pupil Navigation:** on
Randomised: off

17. Riley is investigating how easy it is to scratch different rocks. What property is he investigating?

1/4

- buoyancy
- permeability
- **hardness**
- texture

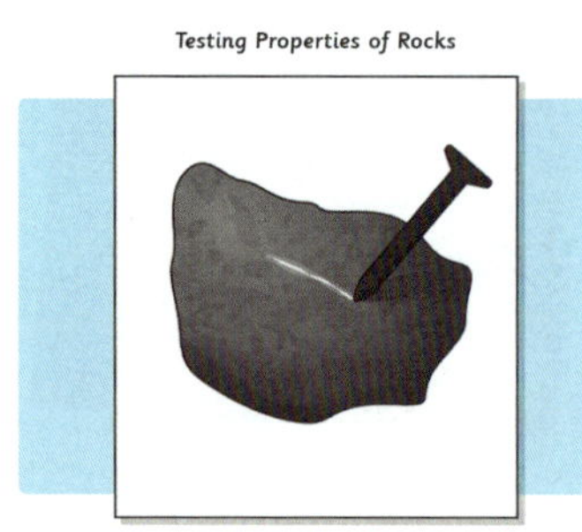

Level 3 *continued*

18. a b c Harrogate, Buxton and Bath are all British spa towns that are known for their spring water. Spring water has lots of dissolved _ _ _ _ _ _ _ _ in it because it is filtered through permeable rocks.

- **minerals**
- **minarals**
- **mineral**
- **minerels**
- **minrals**

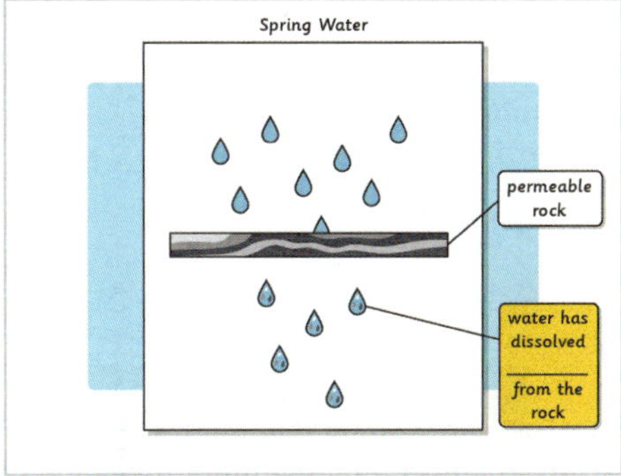

19. a b c Durdle Door is a place in Dorset where the sea has worn a hole in the limestone rock. What property of limestone has allowed the water to wear it away? Explain your answer.

- Open question, no set answer

Durdle Door by Paul Tomlin

Rocks and Soils Vocabulary

Objective: I can use vocabulary relating to rocks and soils.

Quick Search Ref: 10626

Level 1: Knowledge: I can spell and define scientific words for rocks and soils vocabulary.

Required: 10/10 **Pupil Navigation:** on **Randomised:** off

1. A mineral is a ________ substance that makes up rock.

1/3

- man-made
- metal
- natural

minerals by Linda Splashett

2. _ _ _ _ _ _ _ are the ancient remains of plants and animals. They are found in rocks and other substances.

- fossles
- fossils
- fossil
- fosils
- fossuls
- fosles
- fosuls

Triceratops by Tyrell Museum, cinroid by US Government.

3. A _ _ _ _ is made up of different minerals.

- rokc
- rock
- rok
- roc

marble by United States Geological Survey; *quartzite* by Randolph Black.

4. Rocks that water can pass through are. . .

1/4

- impermeable rocks.
- hard rocks.
- permeable rocks.
- soft rocks.

5. Hot, liquid rock called is called _ _ _ _ _ when it has erupted from a volcano.

- lavar
- lava
- magma
- laver

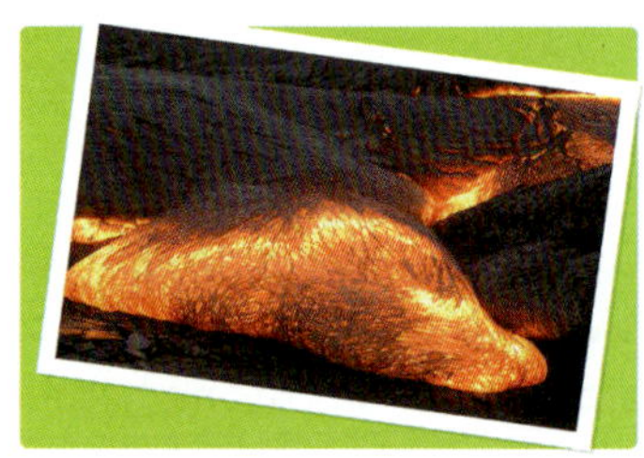

image by Hawaii Volcano Observatory.

6. When hot, liquid rock (lava) cools, it forms different types of __________ rock.

1/3

- sedimentary
- igneous
- metamorphic

7. Sedimentary rocks are formed . . .

1/3

- when magma cools.
- from other types of rocks that are changed due to heat or pressure underground.
- from bits of sand, shells and pebbles.

8. _ _ _ _ is a mixture of water, small pieces of rock and dead plants and animals.

- soyl
- soial
- soil
- soyle
- soyal

9. Select the statement that is true.

1/4

- Impermeable rocks are hard.
- Water can pass through impermeable rocks.
- Impermeable rocks are soft.
- Water can't pass through impermeable rocks.

10. Which of these are rocks?

3/5

- slate
- peat
- chalk
- marble
- loam

Primary Science Y3/4

Forces

Contact and Non-Contact
Magnets

Forces

Objective: I know that some forces need contact between two objects. I can compare how things move on different surfaces.

Quick Search Ref: 10894

Level 1: Knowledge: I can explain how some forces require contact and some do not, giving examples.

Required: 8/10 **Pupil Navigation:** on **Randomised:** off

1. In science, a force is . . .

1/3

- a push.
- a pull.
- **a push or a pull.**

2. Select the three **push** forces.

3/5

- **riding a scooter**
- closing a blind
- **ringing a doorbell**
- **switching on a light**
- peeling an orange

3. Select the three **pull** forces.

3/5

- **pulling on a boot**
- hammering a nail
- closing a drawer
- **opening a drawer**
- **a fridge magnet on a fridge**

4. Which three of these statements about forces are true?

3/4

- **A force can make an object start moving.**
- A force must touch an object to make it move.
- **A force can stop a moving object.**
- **A force can make a moving object change direction.**

5. Select two ways you could use a force to slow down the runaway pram shown in the picture.

2/4

- A. Push the pram down the hill.
- **B. Pull the pram up the hill.**
- C. Pull the pram down the hill.
- **D. Push the pram up the hill.**

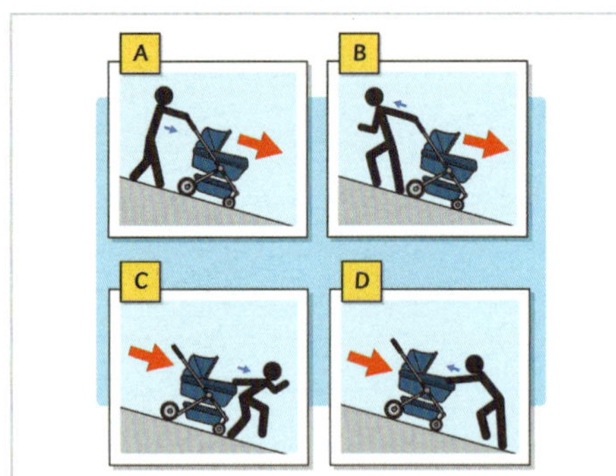

6. What is a non-contact force?

1/2

- a force that needs to touch the thing it's acting on
- **a force that can act on an object without touching it**

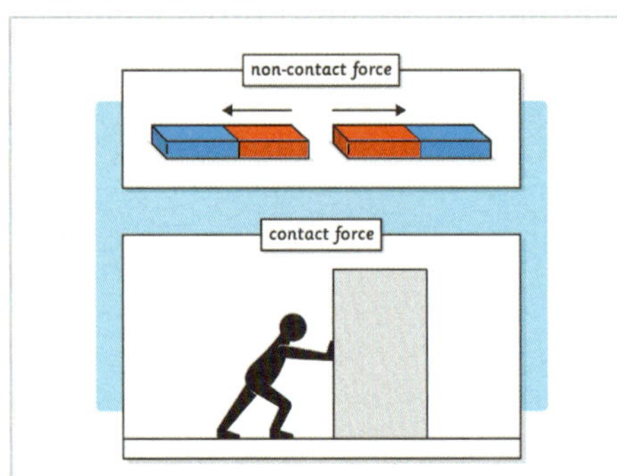

Level 1 *continued*

7. Which two pictures show a force changing the shape of an object?

1/4

- A and B
- B and D
- A and C
- C and D

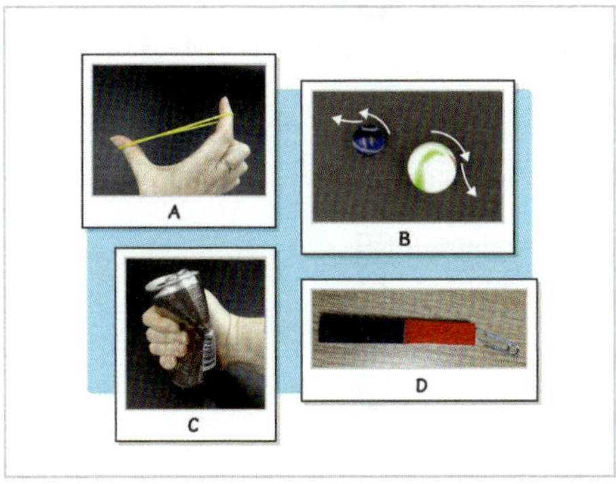

8. Which **two statements** about magnets are true?

2/4

- A magnet only attracts a steel paperclip when the magnet and paperclip are touching.
- When a fridge magnet holds paper on a fridge door, the magnet isn't touching the magnetic material.
- A magnet can attract a steel nail without touching it.
- Magnets always pull other magnets.

9. Contact forces . . .

1/3

- can only act on objects if they are touching them.
- are called friction.
- always push the things they are acting on.

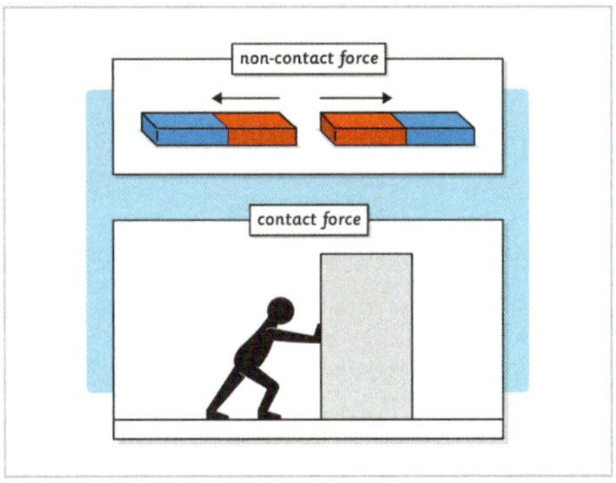

10. __________ is a non-contact force between magnets and magnetic materials or other magnets.

a b c

- friction
- magnetism
- magnetisum
- magnetizm
- magnetizum

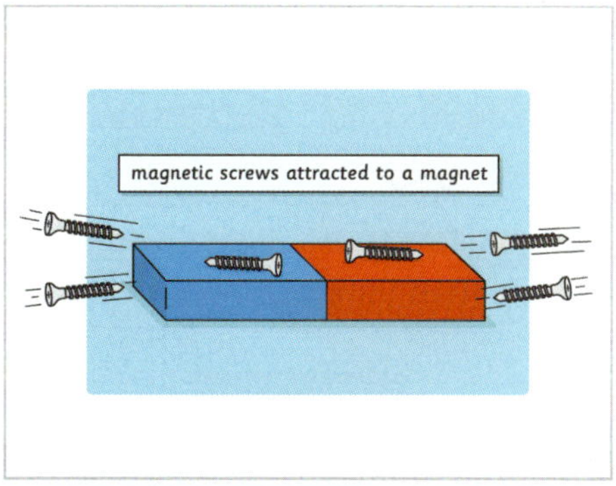

Level 2: Understanding: I can explain why objects move differently on different surfaces.

Required: 5/7 **Pupil Navigation:** on

Randomised: off

11. ________ is a force that slows down moving objects.

a b c

- friction
- fricshion
- gravity
- fricktion
- magnetism
- fricsion
- fricshun
- water resistance
- air resistance

12. A marble rolling on different surfaces will experience different amounts of friction. The more friction that is created between the marble and the surface . . .

1/2

- the faster the marble will roll.
- the slower the marble will roll.

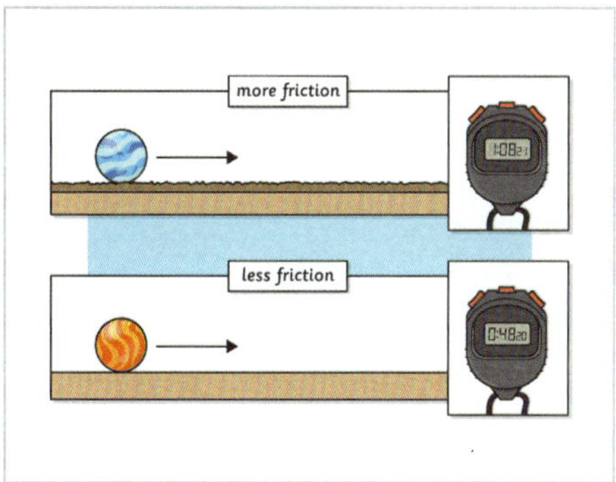

Level 2 *continued*

13. Which three surface textures will create a lot of friction when an object moves against it?

3/5

- smooth
- **sticky** ✓
- **rough** ✓
- oily
- **fluffy** ✓

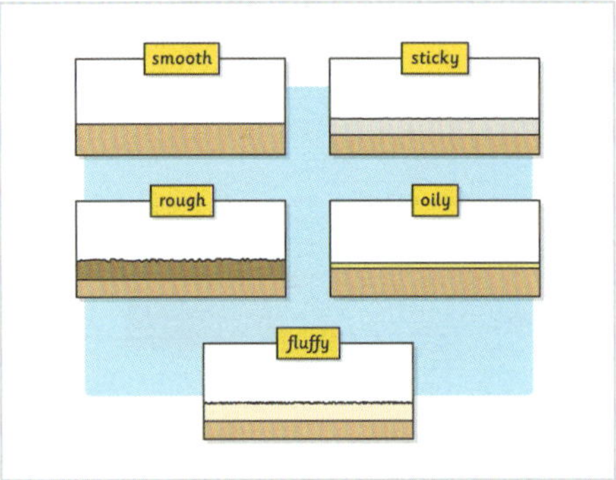

14. Friction acts . . .

1/4

- **in the opposite direction to a moving object.** ✓
- towards the centre of the Earth.
- in the same direction as a moving object.
- away from the centre of the Earth.

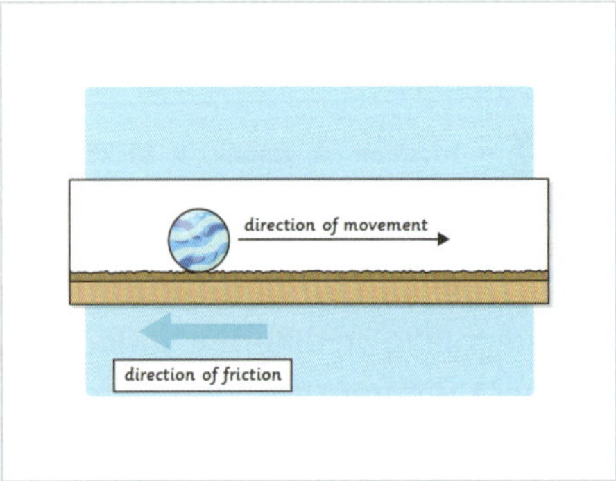

15. Which of these surfaces will create very little friction when an object moves against it?

2/4

- bumpy surface
- **smooth surface** ✓
- jagged surface
- **polished surface** ✓

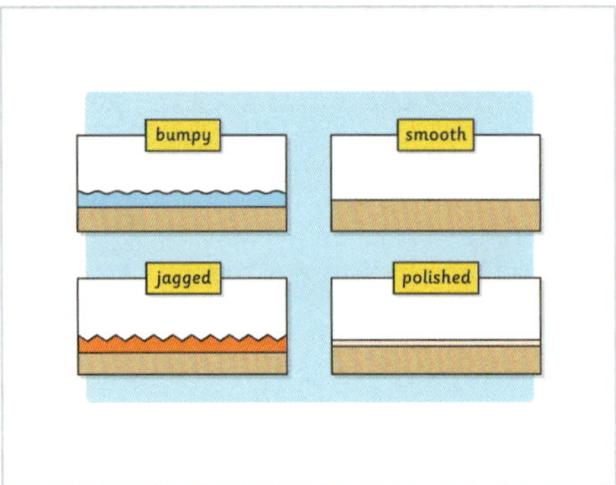

16. Friction is a force that . . .

1/3

- speeds up moving objects.
- attracts objects to a magnets.
- **slows down moving objects.** ✓

17. Which two surface textures will create the least friction when an object moves against it?

2/5

- **smooth** ✓
- sticky
- rough
- **oily** ✓
- fluffy

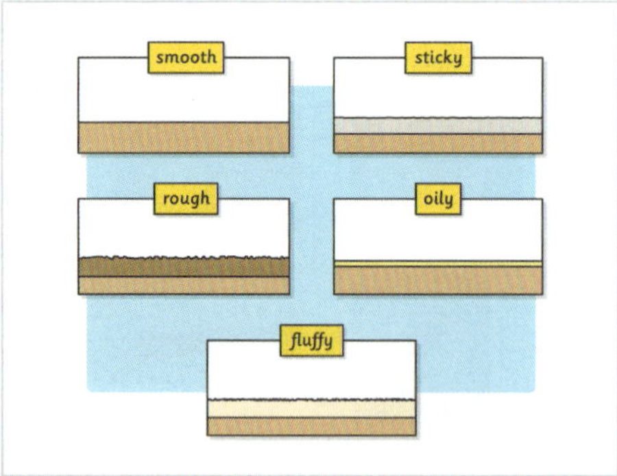

Level 3: Working scientifically to investigate forces: I can make careful and accurate observations and use standard units.

Required: 3/3 **Pupil Navigation:** on
Randomised: off

18. Which piece of equipment can you use to measure the size of a force?

1/4

- a measuring cylinder
- a stopwatch
- **a forcemeter** ✓
- a ruler

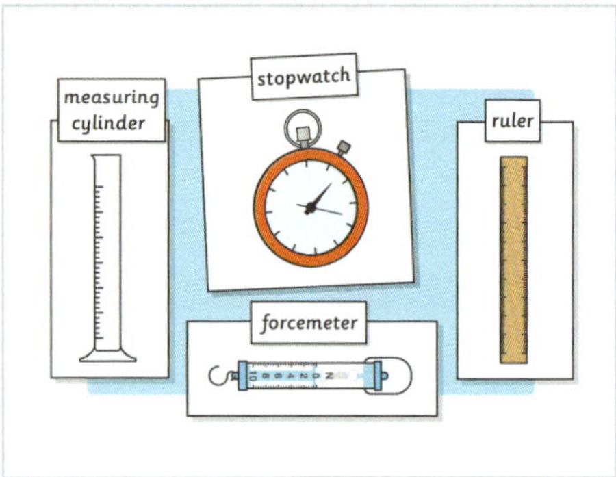

19. What unit are forces measured in?

1/3

- metre (m)
- litre (l)
- **newton (N)** ✓

Level 3 *continued*

20. How much force was needed to open the drawer shown in the diagram?

1/4

- 2 N
- **6 N**
- 10 N
- 25 N

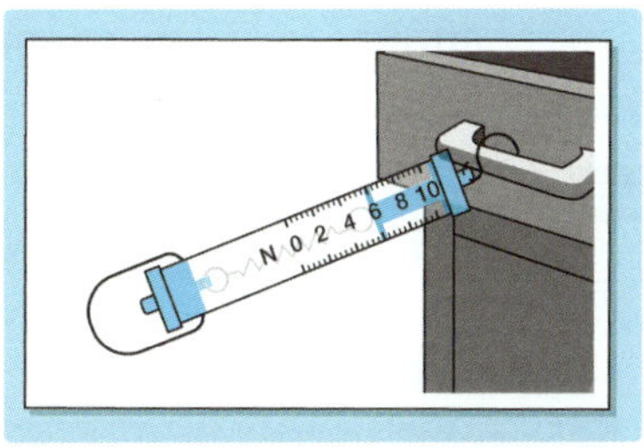

Level 4: Extended reasoning about forces: I can apply my knowledge and understanding of forces to the world around me.

Required: 2/2 **Pupil Navigation:** on
Randomised: off

21. Running racetracks are usually covered in a rough, rubbery material. Explain why it would be more difficult for a runner to run on a smooth surface.

- Open question, no set answer

22. Jasmine has made a fishing game. She has put fish made from magnetic materials into a tank of water and plans to use magnets on the end of fishing rods to attract the fish. Leah says that the game won't work because magnets don't work through water. Is Leah right? Select the statement that explains why Leah is right or wrong.

1/3

- Leah is right. Materials aren't magnetic when they're in water.
- Leah is right. Magnets can only attract things they are touching.
- **Leah is wrong. Magnets work in water and can attract things they aren't touching.**

Forces Vocabulary

Objective: I can use vocabulary relating to forces.

Quick Search Ref: 11040

Level 1: Knowledge: I can spell and define scientific words for forces.

Required: 12/12 **Pupil Navigation:** on **Randomised:** off

1. A _ _ _ _ _ can either be a push or a pull.

abc ■ force

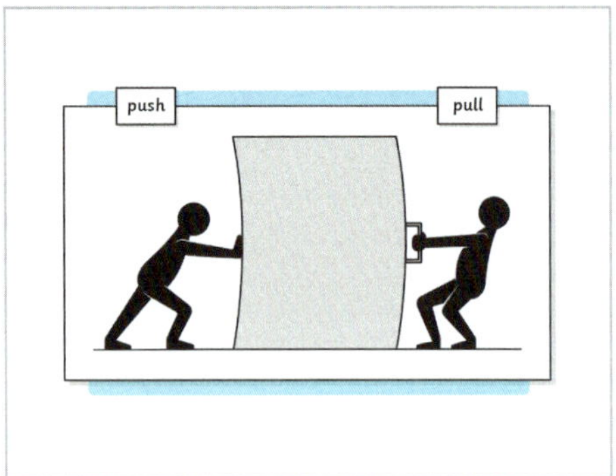

2. If you want to move something away from you, you need to apply a _ _ _ _ force.

abc ■ push ■ pull

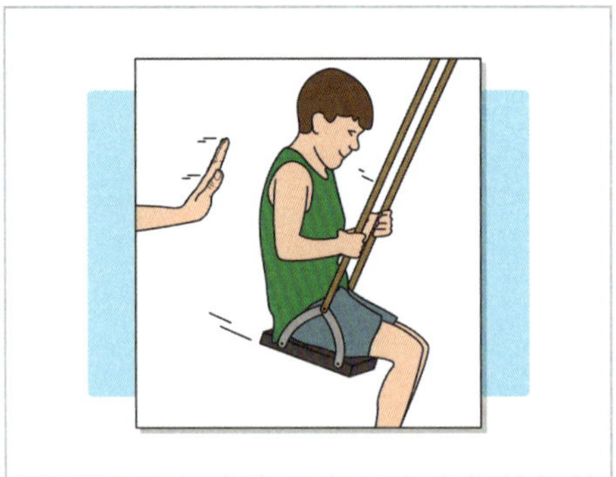

3. _ _ _ _ _ _ _ _ is a force that will slow down a ball rolling across a rough surface.

abc ■ friction

4. If you want to move something towards you, you need to apply a _ _ _ _ force.

abc ■ pull ■ push

5. A _ _ _ _ _ _ attracts (pulls) magnetic materials towards it.

abc ■ magnet

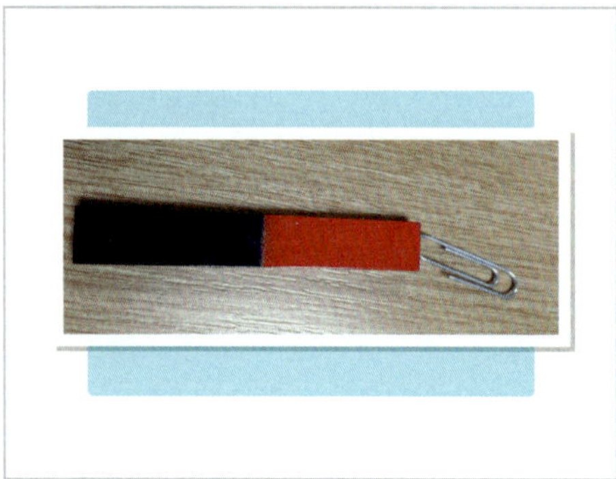

6. When an object applies a force to another object **without touching it**, the force is a ___________ force.

1/2 ■ contact ■ non-contact

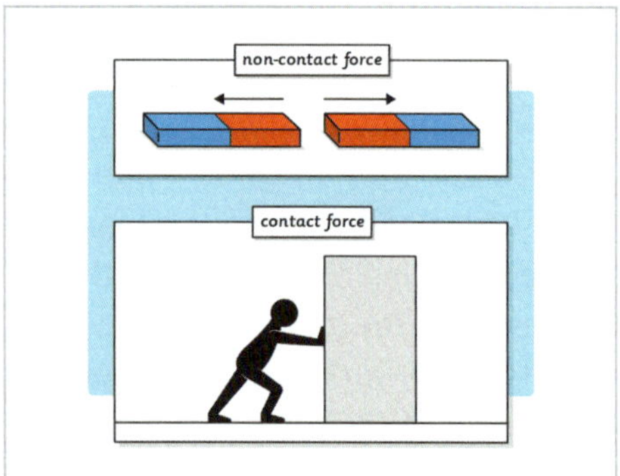

7. North and south are the names for the opposite _ _ _ _ _ of a magnet.

abc ■ pole ■ poles

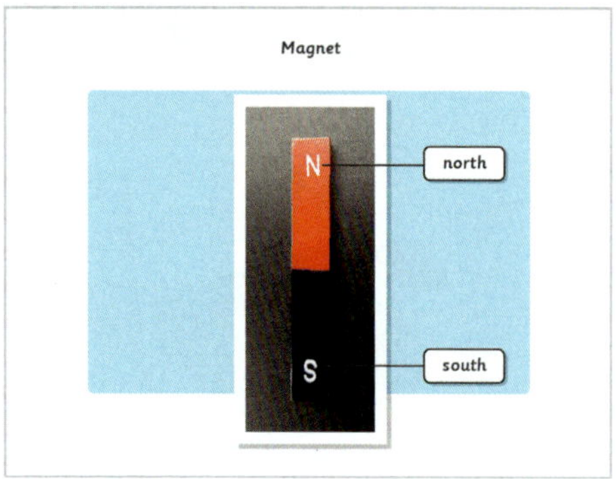

Level 1 *continued*

8. The south pole of one magnet and the north pole of another magnet _ _ _ _ _ _ _ each other. This means that the two magnets will be pulled towards each other.

abc

- attracts
- attract
- repel
- atract

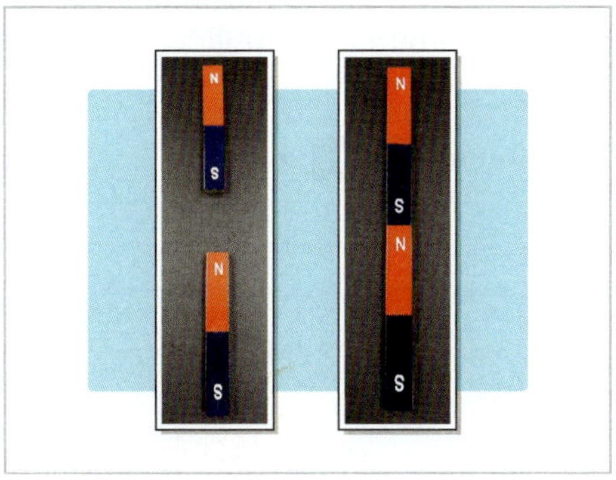

9. When an object applies a force to another object that it's **touching**, the force is a ___________ force.

abc

- contact
- noncontact
- non contact
- non-contact

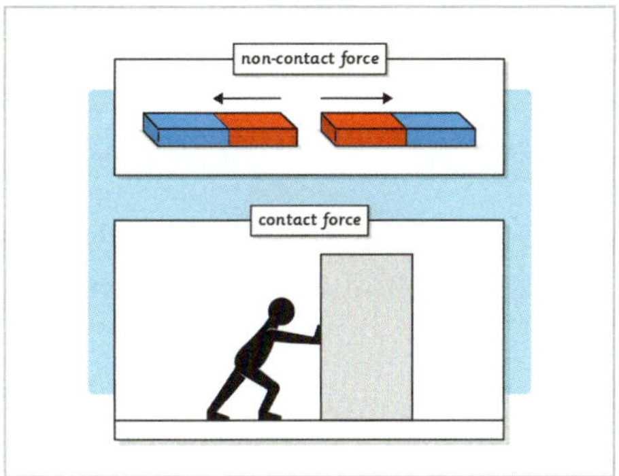

10. Which piece of equipment do you use to measure the size of a force?

1/3

- stopwatch
- force meter
- measuring cylinder

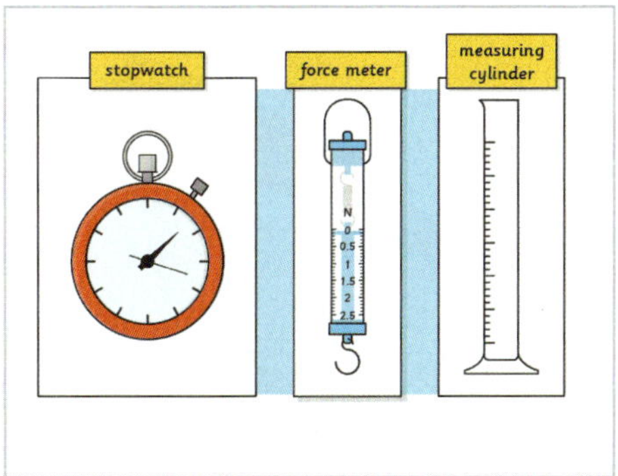

11. The south pole of one magnet and the south pole of another magnet will _ _ _ _ _ each other. This means that the two magnets will push away from each other.

abc

- repel
- repell
- reppel
- attract

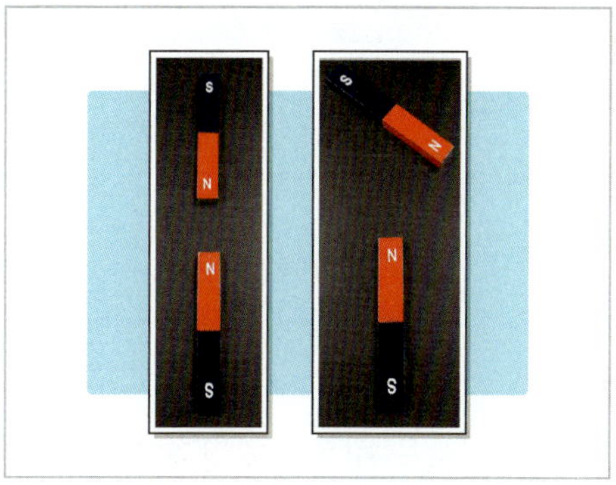

12. The size of a force is measured in . . .

1/4

- metres (m).
- seconds (s).
- newtons (N).
- litres (l).

Magnets

Objective: I know that magnetic forces act without direct contact. I can describe magnets as having two poles that will attract or repel other magnets

Quick Search Ref: 11256

Level 1: Knowledge: I know that magnetism is a non-contact force between magnets and magnetic materials.

Required: 11/14 **Pupil Navigation:** on **Randomised:** off

1. Magnetism is a type of __________.

1/3

- material
- **force**
- friction

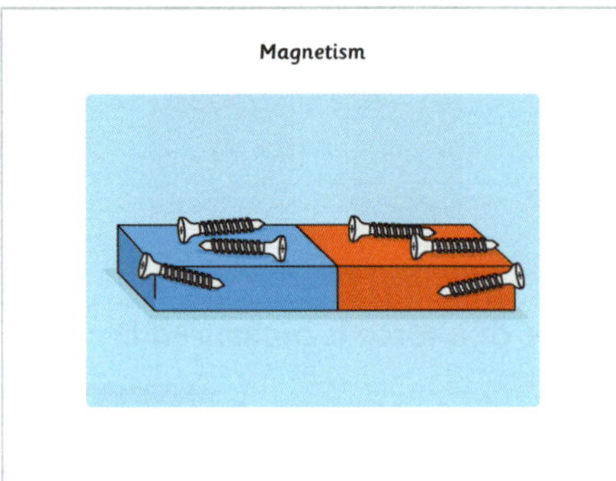

2. Magnetism is a ____________ force.

1/2

- contact
- **non-contact**

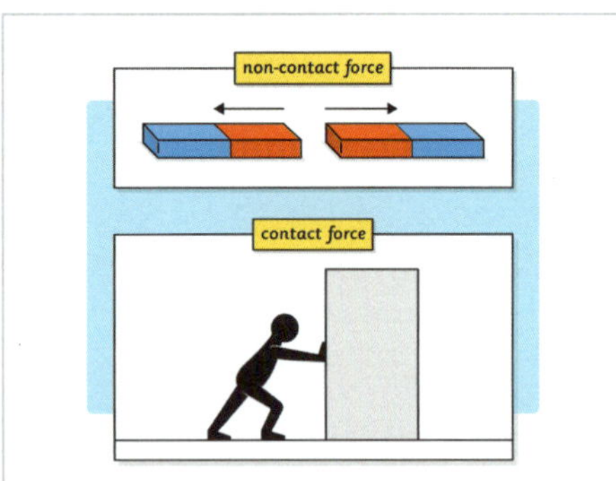

3. A non-contact force . . .

1/2

- needs to be touching the object it's acting on.
- **can act on an object that it isn't touching.**

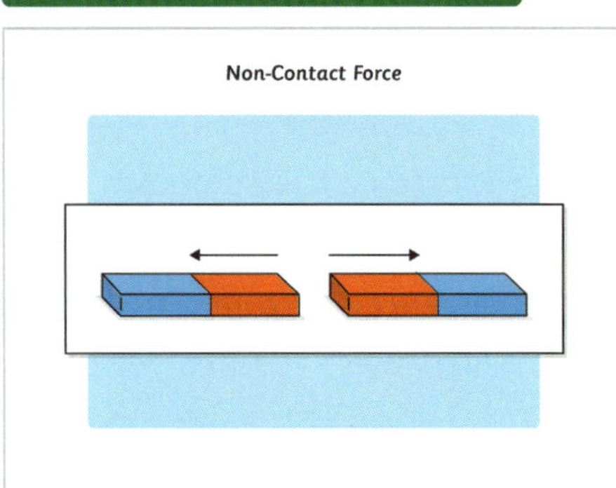

4. A _________ material is pulled towards a magnet.

1/3

- attractive
- **magnetic**
- metallic

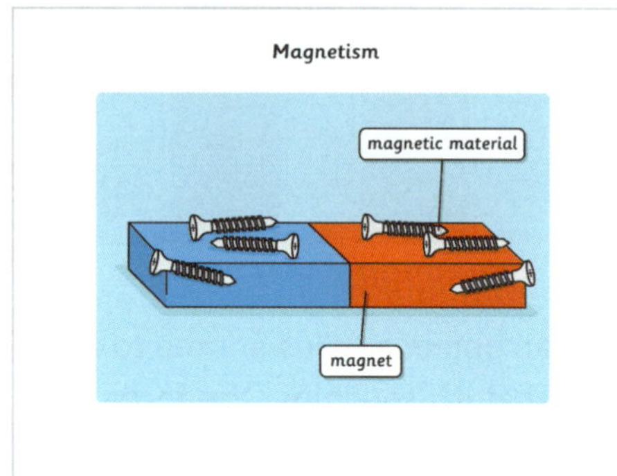

5. Which of these materials is magnetic?

1/5

- copper
- glass
- **iron**
- aluminium
- wood

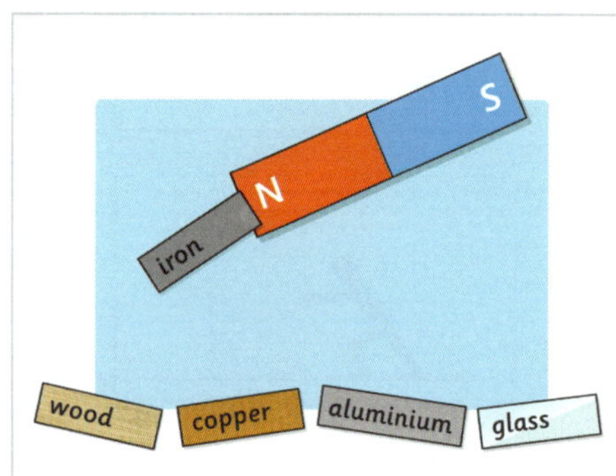

6. Which of these materials are non-magnetic?

4/5

- **copper**
- **glass**
- iron
- **aluminium**
- **wood**

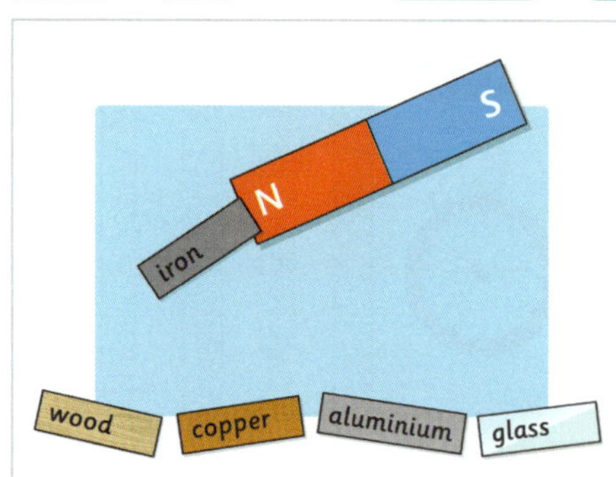

Level 1 *continued*

7. When a magnet pulls a magnetic object towards it, we say that it ________ the object.

1/2

- **attracts**
- repels

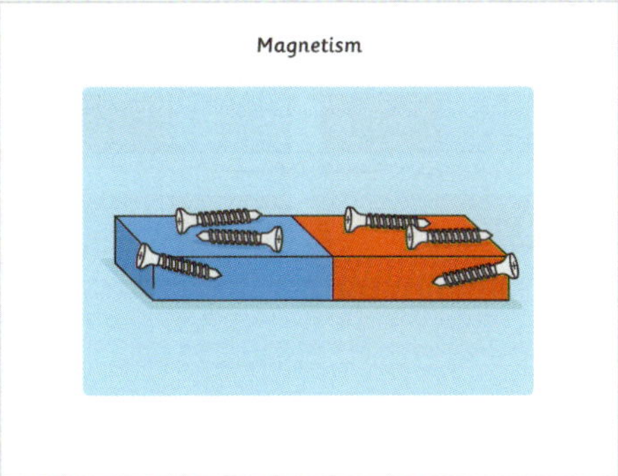

8. The force of magnetism is strongest at the ends of a bar magnet. The strongest points of the magnet are called the magnetic ________.

a b c

- **poles**
- ends
- pole

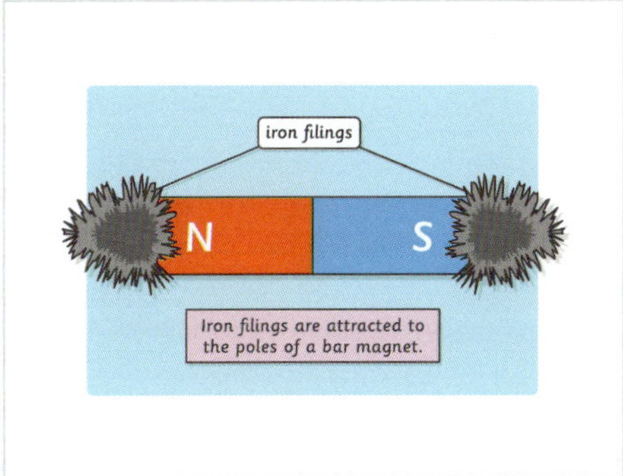

9. All magnets have two magnetic poles. These poles are called . . .

1/3

- top and bottom.
- left and right.
- **north and south.**

10. Two magnets interact when they're brought close together. The pairs of magnets in the diagram are ________ each other.

1/2

- attracting
- **repelling**

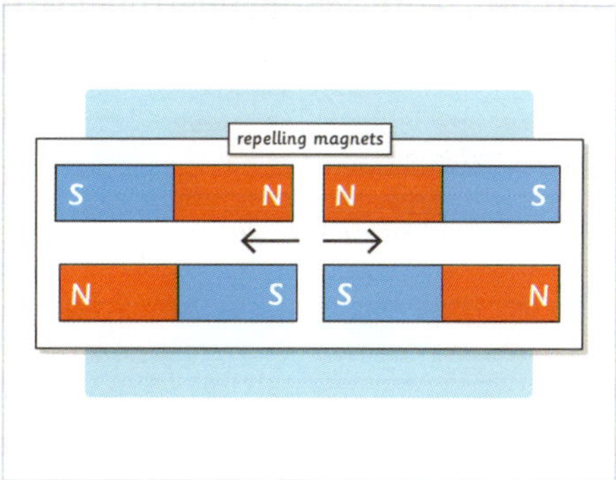

11. When two magnets are put next to each other with the same poles together, they push each other apart. When magnets are pushing apart, we say they are . . .

1/2

- attracting.
- **repelling.**

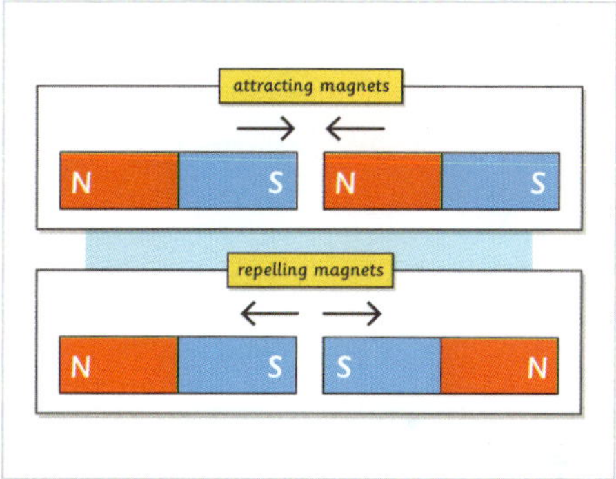

12. Two magnets interact when they're brought close together. The pairs of magnets in the diagram are ________ each other.

1/2

- **attracting**
- repelling

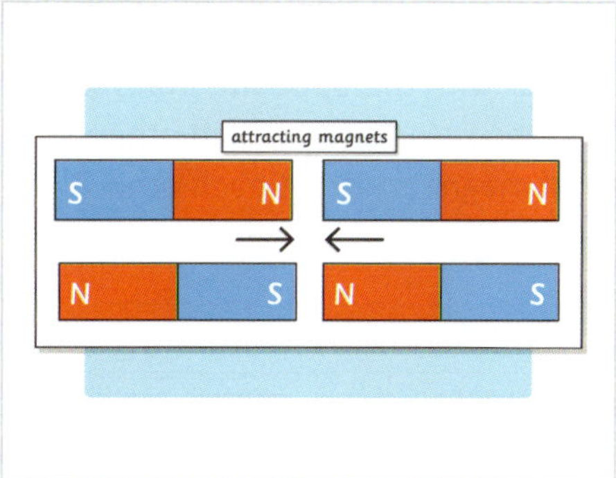

Level 1 *continued*

13. Magnets that are attracting each other . . .

1/2

- **pull together.**
- push apart.

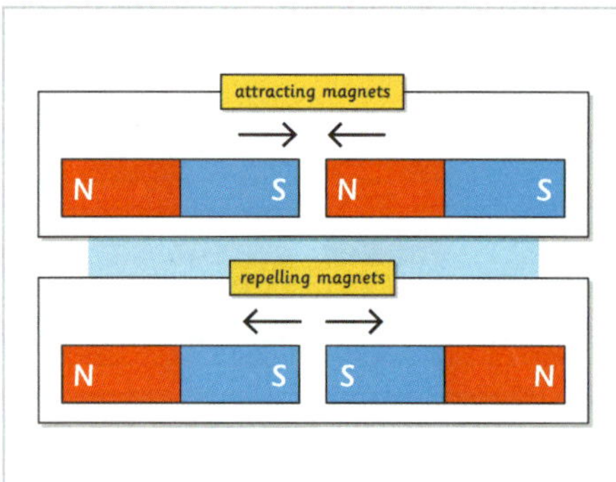

14. What names are given to the poles of a magnet?

1/4

- top and bottom
- east and west
- **north and south**
- red and blue

Level 2: Understanding: I understand how the poles of magnets interact.

Required: 6/8 **Pupil Navigation:** on
Randomised: off

15. A good way to remember how the poles of a magnet affect each other is the saying **opposites attract**. When magnets are placed together with the **opposite** poles facing each other, the poles attract each other. Which pair of magnets attract each other?

1/3

- **A**
- B
- C

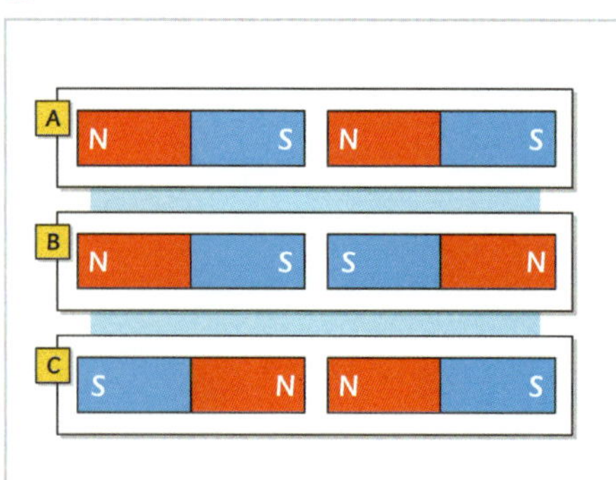

16. When you bring the opposite poles of two magnets together, the poles __________ each other.

1/2

- **attract**
- repel

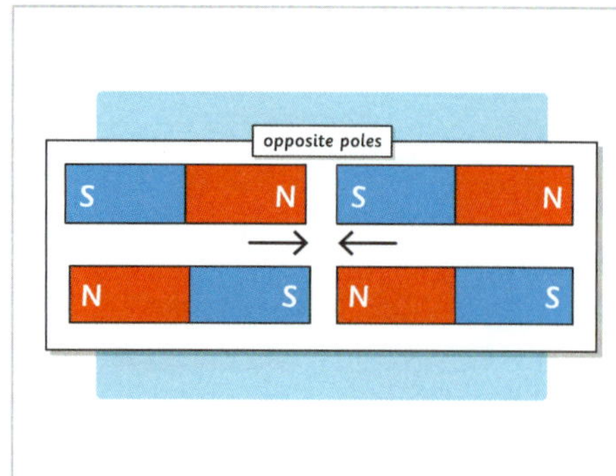

17. When you bring the opposite poles of two magnets together, the poles attract each other. When you bring the like poles of two magnets together, the poles __________ each other.

1/2

- attract
- **repel**

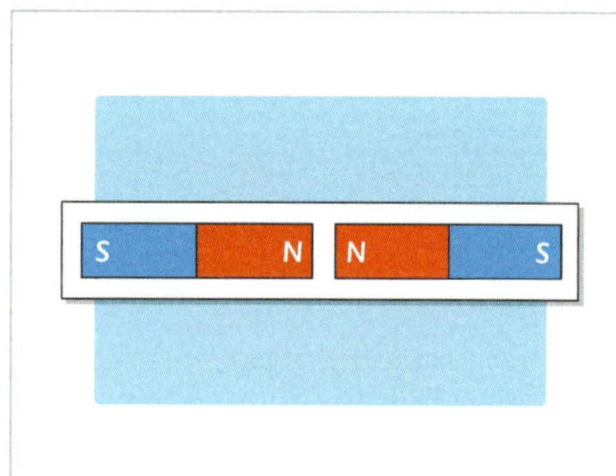

18. When you bring the south poles of two magnets together, the poles __________ each other.

1/2

- attract
- **repel**

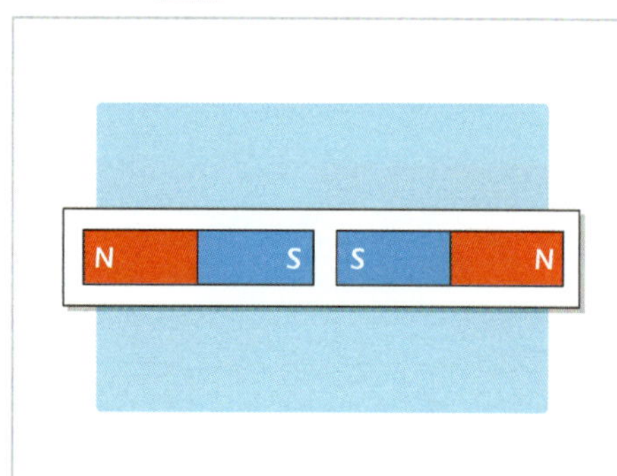

Level 2 *continued*

19. The north pole of one magnet will _______ the south pole of another magnet.

1/2

■ **attract** ■ repel

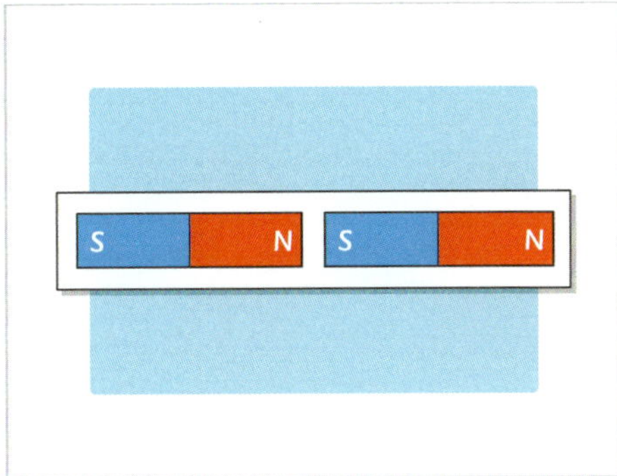

20. When you bring the opposite poles of two magnets together, the poles attract each other. Which two diagrams show magnets that would repel each other?

2/4

■ A ■ **B** ■ **C** ■ D

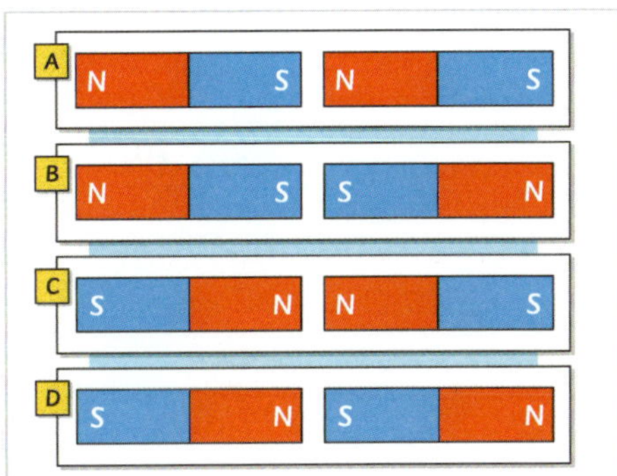

21. If you bring the poles of two magnets together, they either __________ or __________ each other.

2/4

■ pull ■ push ■ **attract** ■ **repel**

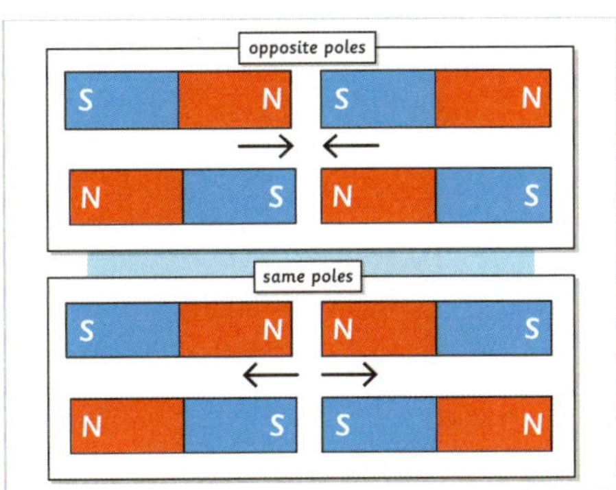

22. Which **two** diagrams show magnets that attract each other?

2/4

■ **A** ■ B ■ C ■ **D**

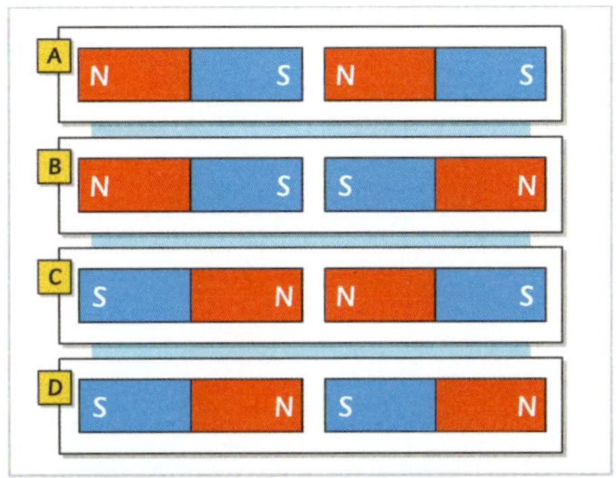

Level 3: Application: I can apply my knowledge and understanding of magnets to real life situations.

Required: 1/1 **Pupil Navigation:** on

Randomised: off

23. The magnetic poles are shown on magnet A. The poles are not shown on magnet B. Describe how you could use magnet A to find out which end of magnet B is the south pole and which is the north pole.

a
b
c

- Open question, no set answer

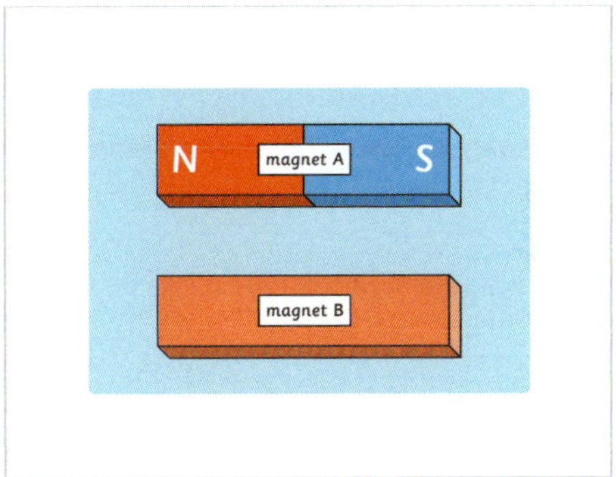

Y3/4

Primary Science

Waves

Light
Sound

Light and Dark

Objective: I can recognise that light is needed to be able to see things and that darkness is the absence of light.

Quick Search Ref: 11252

Level 1: Knowledge: I can describe what light and dark are.

Required: 6/8 | Pupil Navigation: on | Randomised: off

1. We need _ _ _ _ _ to reach our eyes to be able to see.

abc

- **light**
- lite

2. We can't see when it is completely dark, because there is no _______.

abc

- **light**
- lite

3. A light source is something that . . .

1/3

- reflects light.
- **gives out light.**
- we can see.

4. The _ _ _ is the biggest light source in our Solar System.

abc

- **Sun**
- son

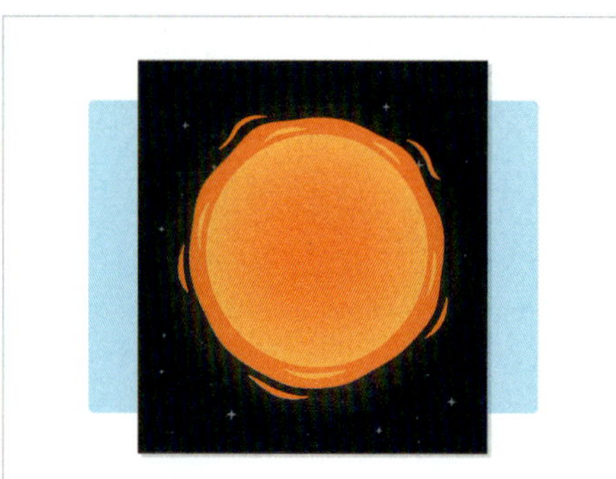

5. If something doesn't give out light, we can only see it when light _ _ _ _ _ _ _ _ off it.

1/2

- **reflects**
- reflect

6. Looking at the Sun is dangerous. Select the statement that is true.

1/3

- You should always wear safety goggles to protect your eyes when you look at the Sun.
- **You should never look directly at the Sun.**
- It's safe to look at the Sun if you are wearing sunglasses.

7. Something that gives out light is called . . .

1/3

- reflective.
- **a light source.**
- visible.

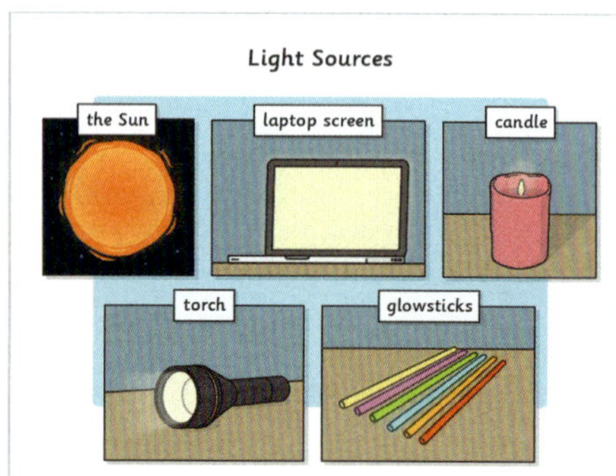

Level 1 *continued*

8. When it is completely dark, there is ____ light.

1/2

- still some
- no

Level 2: Understanding: I can explain that we need light to be able to see.

Required: 6/6 **Pupil Navigation:** on

Randomised: off

9. Which **three** of these things are light sources?

3/5

- the Sun
- a mirror
- a window
- a laptop screen
- glow stick

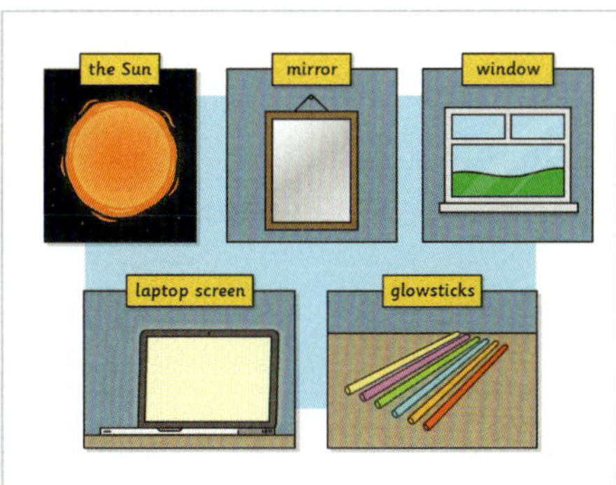

10. Freddie, Lacey and Theo are discussing what will happen if Jenny turns the main light off in their classroom but leaves the lamp on. Who has the right idea?

1/3

- Freddie
- Lacey
- Theo

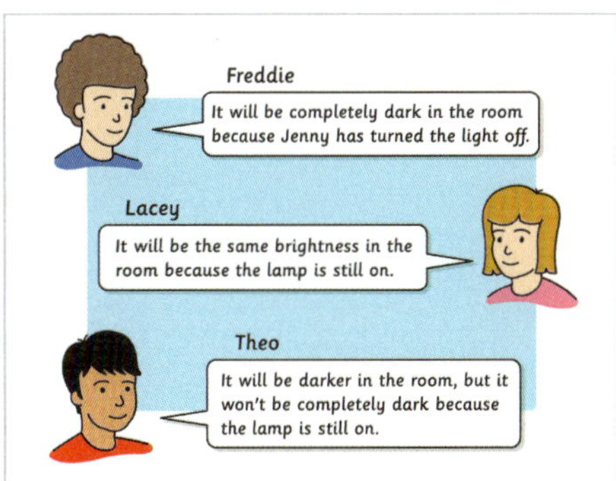

11. Eliza is looking at her laptop screen. For her to be able to see the screen, which direction must the light be travelling?

1/2

- from her eyes to the screen
- from the screen to her eyes

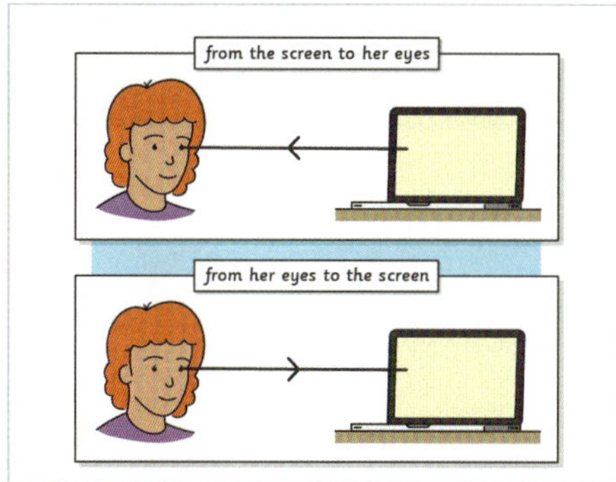

12. How can Esmee see the cupboard in her room?

1/4

- The cupboard is reflecting light from the lamp.
- Light is travelling through the cupboard.
- Light from the Sun is travelling through the window and reflecting off the cupboard.
- The cupboard is giving out light.

13. What is the light source in this room?

1/4

- the ceiling light
- the lamp
- the TV
- the window

Level 2 *continued*

14. Olivia says, "I can see the table in my kitchen even when the lights are off."
How is this possible?

1/3

- Olivia can see in the dark.
- The table is making its own light.
- Light from the Sun is travelling through the window and reflecting off the table.

Level 3: Application: I can apply my knowledge of light and dark to the world around me.

Required: 4/4 **Pupil Navigation:** on
Randomised: off

15. The Moon isn't a light source. It looks like it's giving off light, but it's actually reflecting light from . . .

1/3

- the Sun.
- Earth.
- the sky.

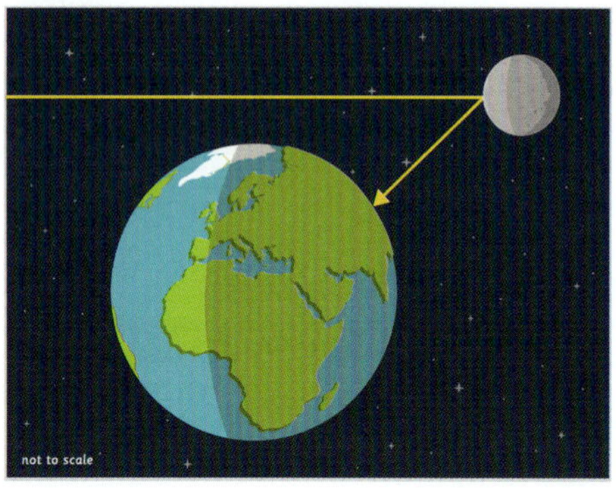

16. A _ _ _ _ _ _ is formed when an object blocks the light.

- shadow

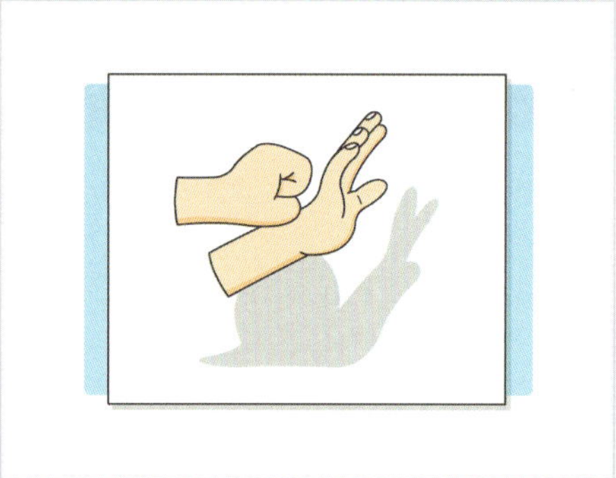

17. Some animals that live deep in the ocean produce light with their bodies. Why do you think this is?

1/3

- because it's dark at the bottom of the ocean
- because there's lots of light at the bottom of the ocean
- because it's cold at the bottom of the ocean

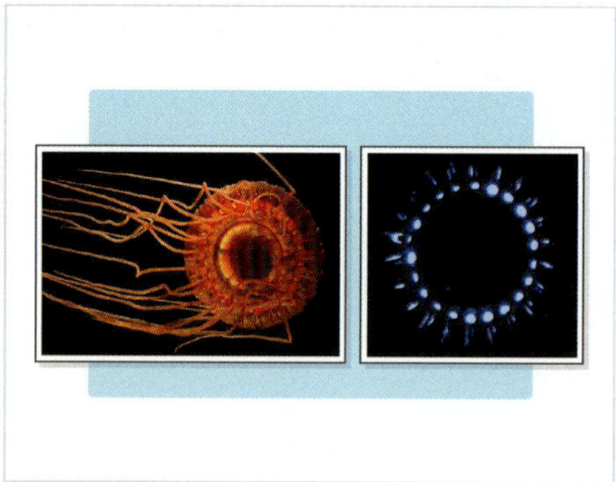

18. When Reba closes her bedroom curtains during the day, her room gets darker. Why is this? Explain your answer.

- Open question, no set answer

Light Vocabulary

Objective: I can use the key words associated with light.

Quick Search Ref: 11262

Level 1:

Required: 9/9 Pupil Navigation: off Randomised: on

1. We need _ _ _ _ _ to reach our eyes to be able to see.

abc

■ light ■ lite

2. We can't see when it is completely dark, because there is no _______.

abc

■ light ■ lite

3. The _ _ _ is the biggest light source in our Solar System.

abc

■ Sun ■ son

4. If something doesn't give out light, we can only see it when light _ _ _ _ _ _ _ _ off it.

1/2

■ reflects ■ reflect

5. A _ _ _ _ _ _ is formed when an object blocks the light.

abc

■ shadow

6. Light can't travel through ________ objects.

abc

■ opaque

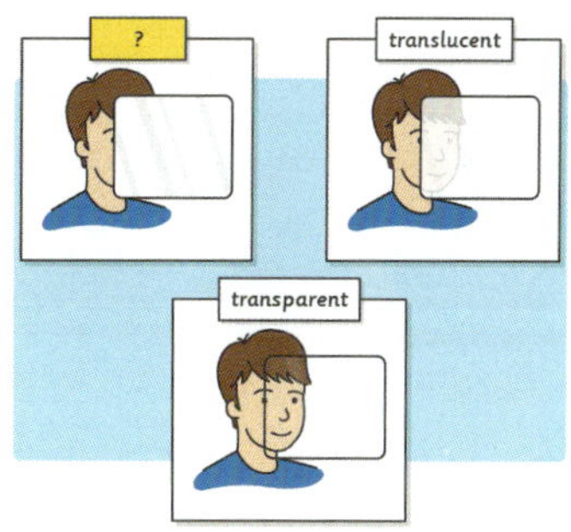

7. You can see clearly through ______________ material because light travels through it easily.

1/3

■ a translucent ■ a transparent ■ an opaque

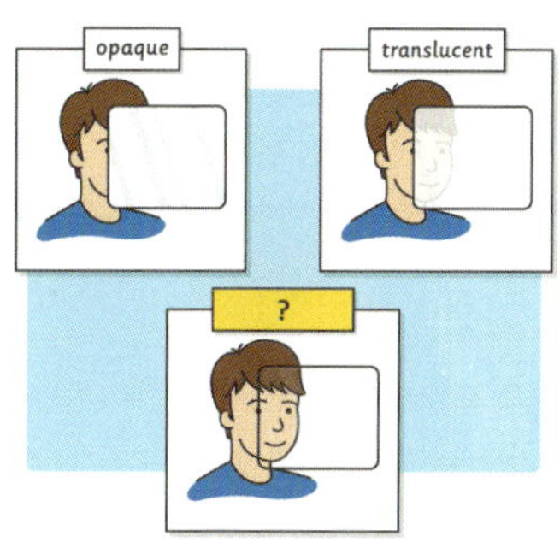

Level 1 *continued*

8. Things usually look blurry when you look through __________ material because only some light can travel through it.

1/3

- a translucent
- a transparent
- an opaque

9. When light can't travel through an object, a __________ forms.

- shadow
- shaddow

Shadows

Objective: I can recognise that shadows are formed when the light from a light source is blocked by an opaque object.

Quick Search Ref: 11253

Level 1: Knowledge: I can explain how a shadow is formed.

Required: 6/7 | Pupil Navigation: on | Randomised: off

1. We need _ _ _ _ _ to reach our eyes to be able to see.

- **light** (correct)
- lite

2. You can see clearly through ______________ material because light travels through it easily.

1/3

- a translucent
- **a transparent** (correct)
- an opaque

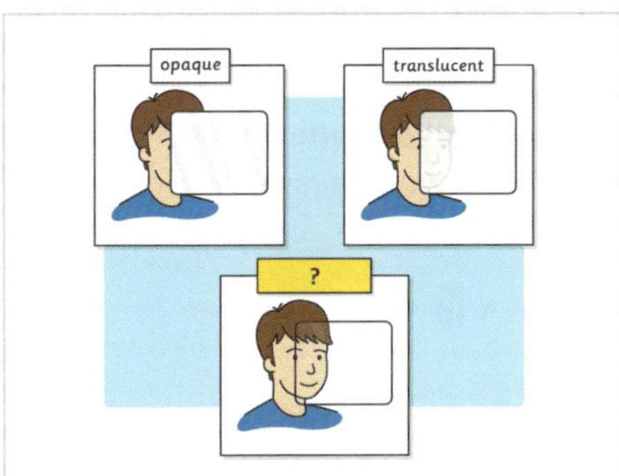

3. Things usually look blurry when you look through __________ material because only some light can travel through it.

1/3

- **a translucent** (correct)
- a transparent
- an opaque

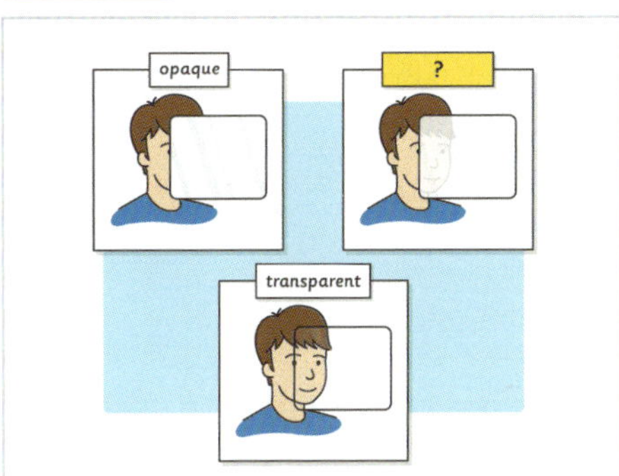

4. Light can't travel through _________ material, so you can't see through it.

1/3

- a translucent
- a transparent
- **an opaque** (correct)

5. When light can't travel through an object, a _________ forms.

- **shadow** (correct)
- shaddow

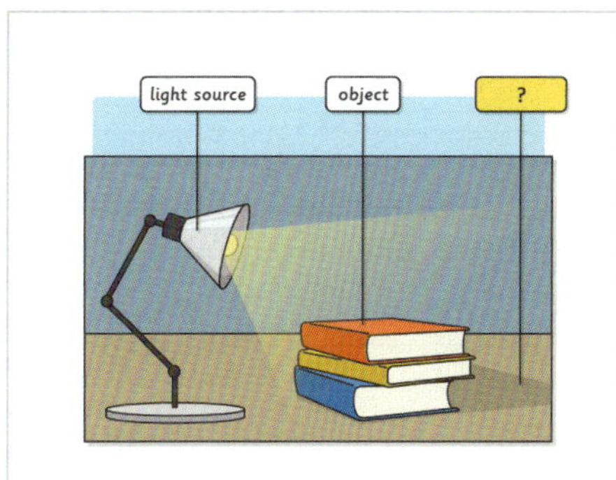

6. What is a shadow?

1/3

- A shadow is the image made when something reflects light.
- **A shadow is a dark area behind an object that light can't pass through.** (correct)
- A shadow is what is produced when something makes its own light.

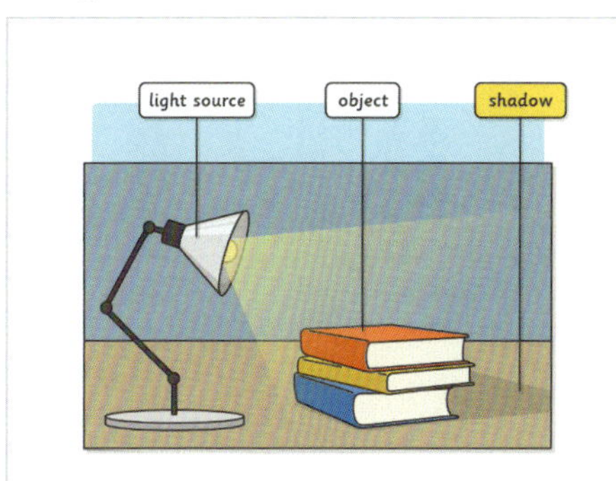

Level 1 *continued*

7. Light can't pass through ________ objects.

1/3

- translucent
- transparent
- **opaque**

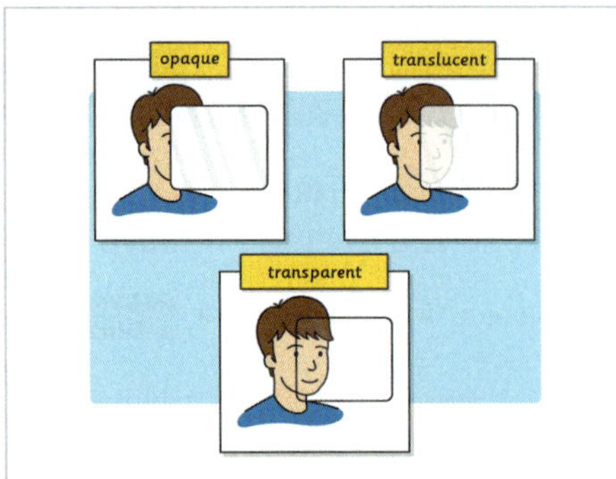

Level 2: Understanding: I can demonstrate how a shadow is formed.

Required: 4/6 **Pupil Navigation:** on

Randomised: off

8. Which image shows the correct shadow for the ball?

1/4

- **A**
- B
- C
- D

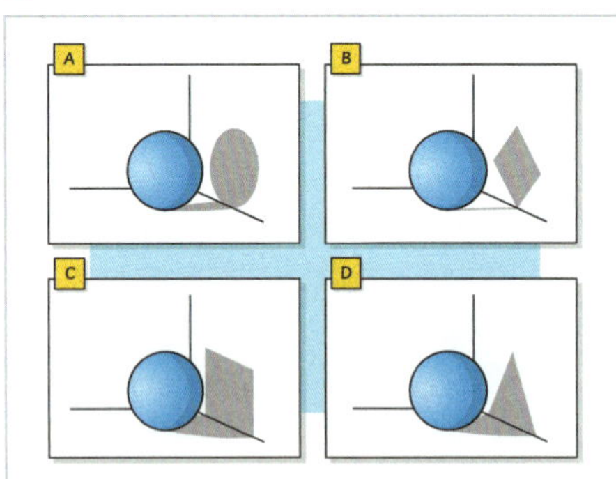

9. The bottle in the diagram is made from blue opaque plastic. Which image shows the correct shadow?

1/4

- A
- B
- **C**
- D

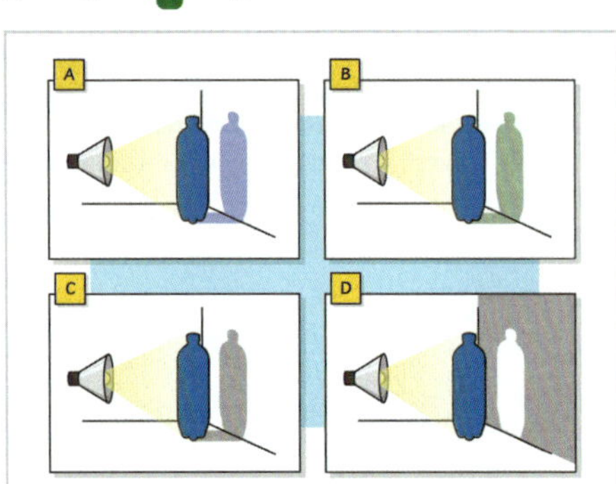

10. Which image shows the correct shadow of the puppet?

1/4

- A
- B
- **C**
- D

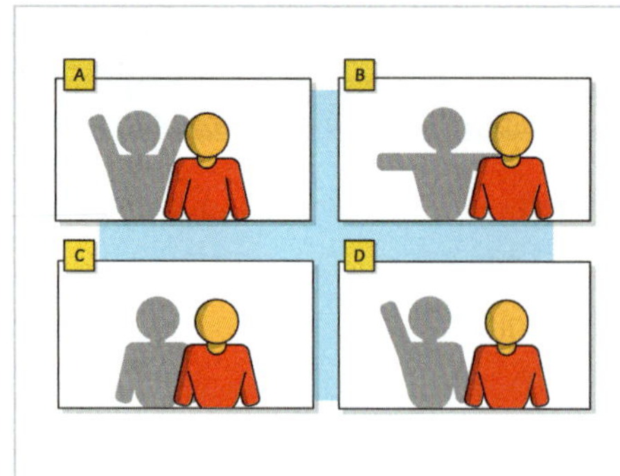

11. Which image shows the correct shadow of Lucy?

1/4

- A
- **B**
- C
- D

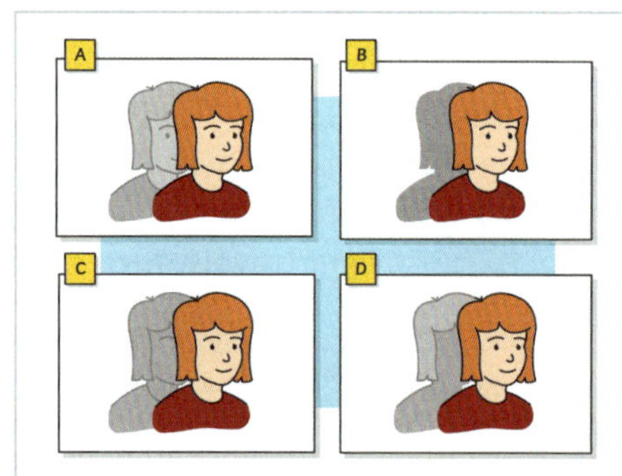

12. The table is in a room where the only light source is the ceiling light. Which shadow is in the correct place?

1/4

- A
- B
- **C**
- D

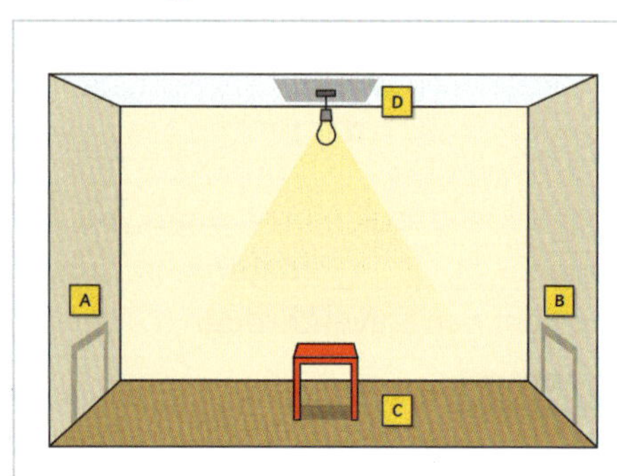

Level 2 *continued*

13. Luca, Ruby and Alfie have noticed that the shadow of a tree in the playground changes size during the day. Who has the right idea about why this is?

1/3

- Luca
- Ruby
- **Alfie**

Level 3: Application: I can apply my knowledge and understanding of shadows to the world around me.

Required: 4/4 **Pupil Navigation:** on
Randomised: off

14. The pointer on a sundial casts a shadow on to a clock face. The position of the shadow can be used to tell the time. The children are discussing how a sundial works. Who has the right idea?

1/3

- Isla
- **Leon**
- Gabriel

15. Light from the Sun can damage your skin. How does sitting in the shade stop you from getting burnt?

- Open question, no set answer

16. A lighthouse is used to alert boats to the danger of rocks at night. It shines a bright light over the water for the people on the boats to see. On which boat will the people see the light from the lighthouse brightest?

1/3

- **A**
- B
- C

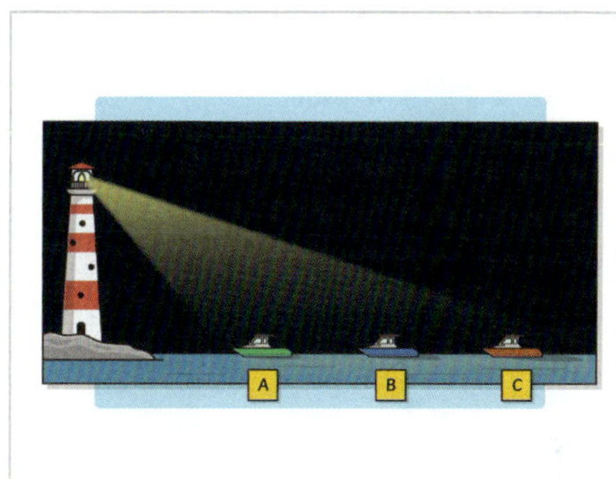

17. A lighthouse is used to alert boats to the danger of rocks at night. It shines a bright light over the water for the people on the boats to see. Explain why a shadow is cast behind each boat.

- Open question, no set answer

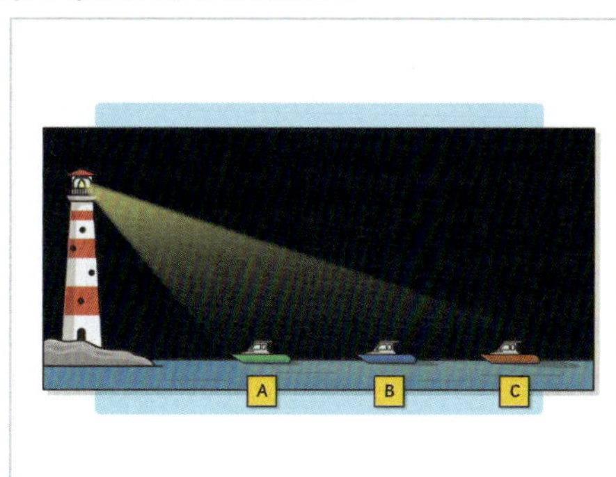

Changing the Pitch of a Sound

Objective: I can understand the relationship between the pitch of a sound and the object producing the sound.

Quick Search Ref: 11257

Level 1: Knowledge: I know the scientific words that are used in the topic of sound.

Required: 6/8 **Pupil Navigation:** on **Randomised:** off

1. A _ _ _ _ _ is made when an object vibrates.

abc
- sound

A Vibrating Tuning Fork

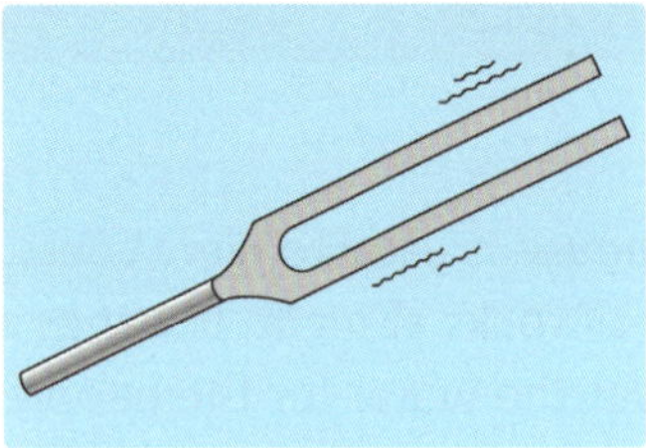

2. The _ _ _ _ _ of a sound describes how high or low the sound is.

abc
- volume
- pitch
- pich

Pitch

3. The _ _ _ _ _ _ of a sound describes how loud the sound is.

abc
- volume
- volyume
- pitch

Volume

4. You hear sound because __________ waves travel from the sound source to your ear.

1/3
- sound
- light
- heat

A Sound Wave Travelling Between an Alarm Clock and an Ear

5. When an object vibrates more ________, it makes a higher-pitched sound.

1/2
- quickly
- slowly

Pitch

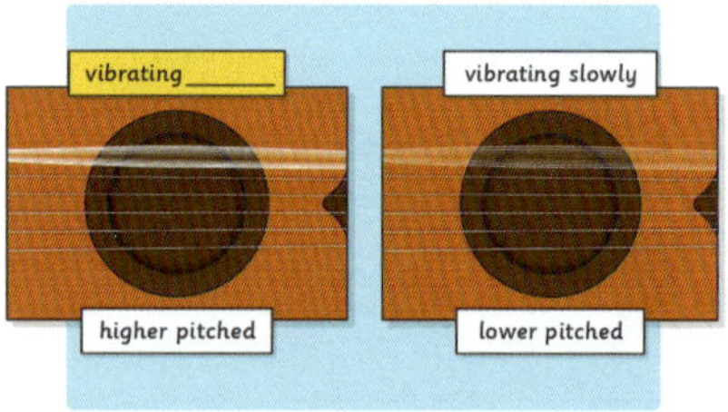

6. Small animals, such as mice, mostly make sounds that are . . .

1/2
- low pitched.
- high pitched.

Field Mouse

field mouse by TJerzystrzelecki

Level 1 *continued*

7. Large animals, such as elephants, mostly make _________ sounds.

1/2

- **low pitched**
- high pitched

8. When an object vibrates more ________, it makes a lower-pitched sound.

1/2

- quickly
- **slowly**

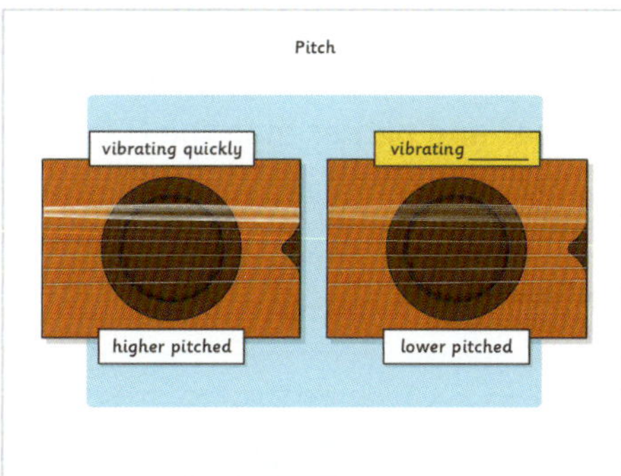

Level 2: Understanding: I can describe how to change the pitch of some musical instruments.

Required: 12/12 **Pupil Navigation:** on
Randomised: off

9. The _ _ _ _ _ _ _ of a guitar vibrate when they are plucked.

a b c

- **strings**

10. The pitch of a note made by a guitar depends on the length, __________ and tightness of the string.

1/3

- position
- colour
- **thickness**

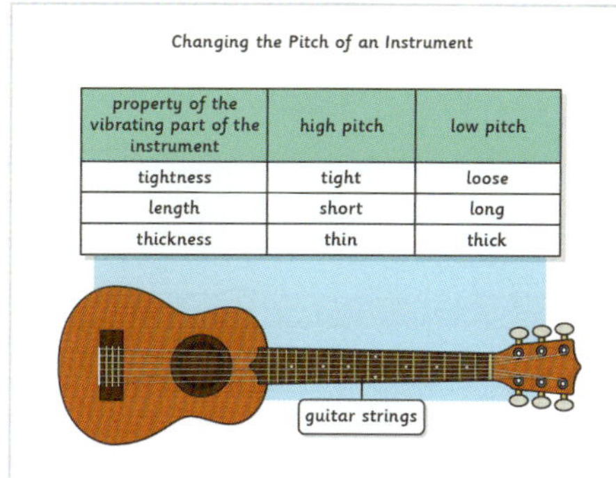

11. The pitch of a note made by a guitar depends on the length, thickness and tightness of the string. A _________, _________ or _______ string makes a higher-pitched sound.
Select three words to complete the statement correctly.

3/6

- **shorter**
- **tighter**
- longer
- looser
- **thinner**
- thicker

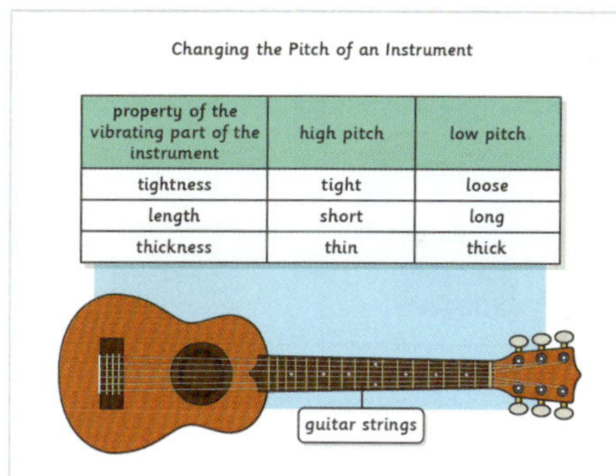

12. When you make the skin of a drum tighter, the pitch of the note made by the drum gets _________.

1/2

- **higher**
- lower

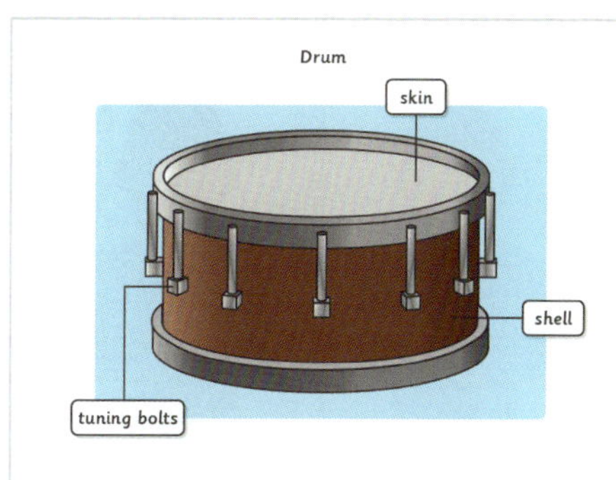

Level 2 *continued*

13. A recorder is a wind instrument that contains a column of ____ that vibrates to make a sound.

a b c

- **air**

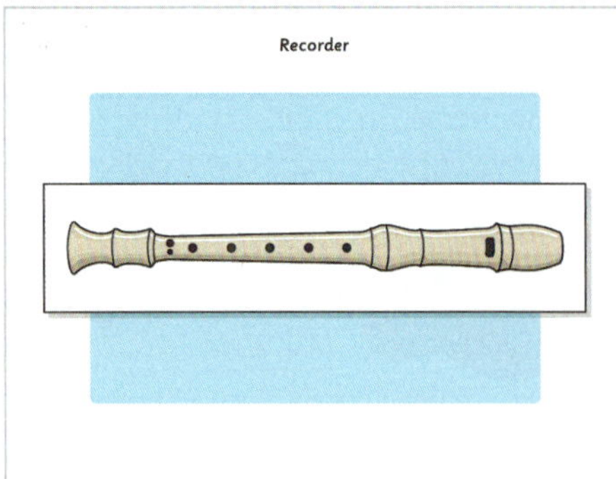

14. The pitch of a note made by a recorder depends on the ______ of the column of air. This can be changed by covering the holes with your fingers as you are blowing.

1/2

- **length**
- diameter

15. The pitch of a note made by a recorder depends on the length of the column of air inside the recorder. Covering the holes on the recorder makes the vibrating column of air _______.

1/2

- shorter
- **longer**

16. The pitch of a note made by a recorder depends on the length of the column of air inside the recorder. Covering the holes on the recorder makes the vibrating column of air longer. A longer column of air gives a _________ sound.

1/2

- higher-pitched
- **lower-pitched**

17. The strings of a violin are usually tuned to play the musical notes G, D, A and E. The strings are the same length, but the E string is thinner and tighter than the G string, so the E string makes a ___________ sound than the G string.

1/2

- **higher-pitched**
- lower-pitched

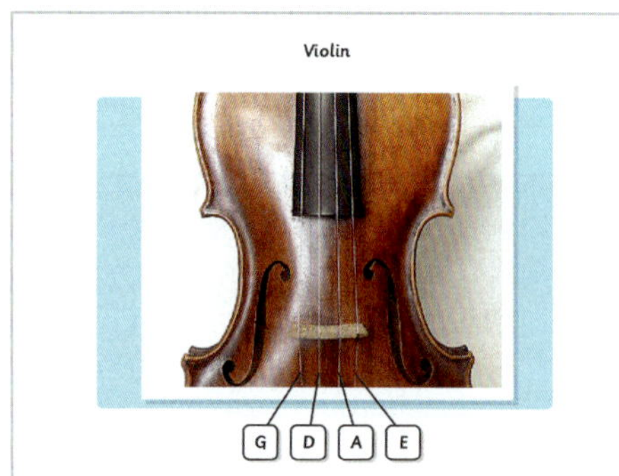

Level 2 *continued*

18. Faith notices that if she presses down on a guitar string, the pitch of the string changes. As she moves her finger along the string towards the body of the guitar, the vibrating part of the string gets shorter and the pitch . . .

1/2

- gets lower.
- **gets higher.**

19. When you hit a drum, the _____ vibrates and makes a sound.

- **skin**

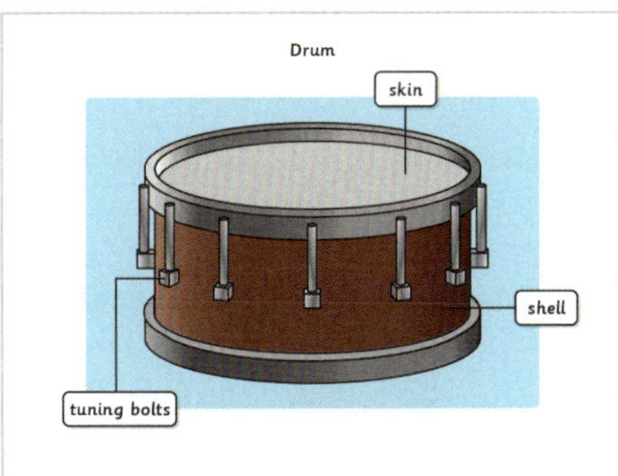

20. The pitch of a note made by a string instrument can depend on the length of the string. A ________ string makes a lower-pitched sound.

1/2

- **longer**
- shorter

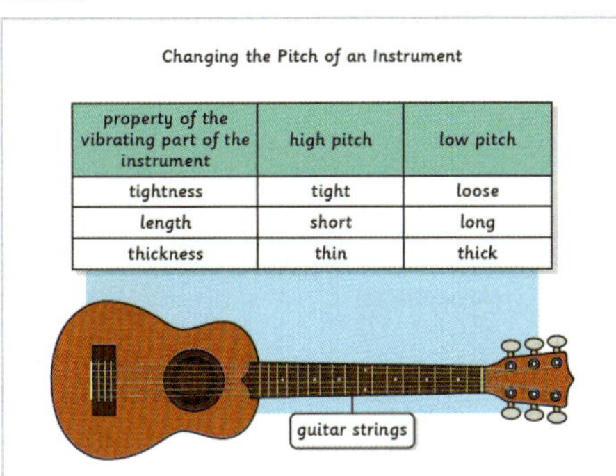
Changing the Pitch of an Instrument

property of the vibrating part of the instrument	high pitch	low pitch
tightness	tight	loose
length	short	long
thickness	thin	thick

Level 3: Application: I can link my knowledge and understanding of sound and pitch to new scenarios.

Required: 1/1 **Pupil Navigation:** on
Randomised: off

21. This photograph shows 'The Singing Ringing Tree' sculpture. It stands on a windy hillside in Lancashire and is made from pipes of different lengths, arranged in different ways. Using your understanding of sound, explain why the sculpture makes sounds with different pitches.

- Open question, no set answer

Singing Ringing Tree by Wendy Mann and Henry Brett

Changing the Volume of a Sound

Objective: I can understand the relationship between the volume of a sound and the strength of the vibrations that produced it.

Quick Search Ref: 11272

Level 1: Knowledge: I know the key information and the scientific words that are used in the topic of sound.

Required: 6/6 **Pupil Navigation:** on **Randomised:** off

1. A _ _ _ _ _ is made when an object vibrates.

- **sound**

A Vibrating Tuning Fork

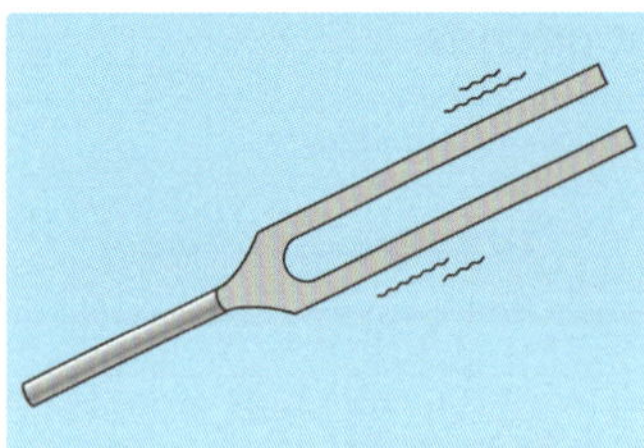

2. The _ _ _ _ _ of a sound describes how high or low the sound is.

- **pitch**
- volume
- pich

Pitch

3. The volume of a sound describes how _____ the sound is.

1/3

- high
- **loud**
- low

Volume

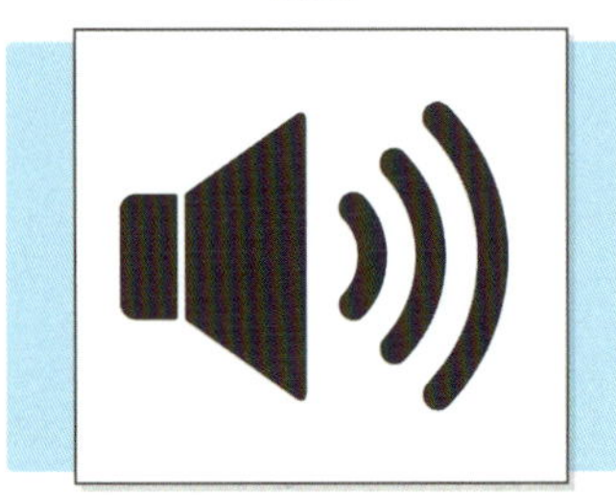

4. You hear a sound when sound _ _ _ _ _ travel from the sound source to your ear.

1/2

- **waves**
- vibrations

A Sound Wave Travelling Between an Alarm Clock and an Ear

5. When you put your hand on your throat while you're talking, you can feel your vocal cords vibrate to make the sound of your _______.

- **voice**

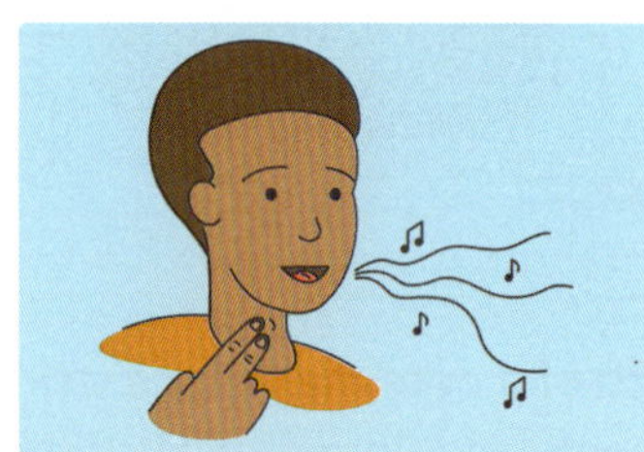

6. What equipment can be used to measure the volume of a sound?

1/3

- **data logger with a sound sensor**
- measuring cylinder
- stopwatch

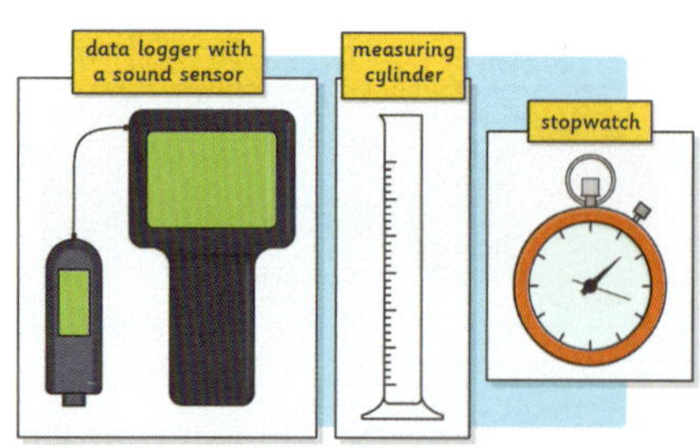

Level 2: Understanding: I can describe how the volume of a sound can be changed.

Required: 6/6 **Pupil Navigation:** on
Randomised: off

7. To make a louder sound on a drum, you need to . . .

1/4

- tighten the drum skin.
- **hit the drum skin harder.**
- loosen the drum skin.
- hit the drum skin more softly.

8. To make a quieter sound on a guitar, you need to . . .

1/4

- tighten the guitar strings.
- pluck the string harder.
- loosen the guitar strings.
- **pluck the guitar string more gently.**

9. Quiet sounds have a . . .

1/4

- high volume.
- low pitch.
- high pitch.
- **low volume.**

10. Loud sounds have a . . .

1/4

- high pitch.
- low volume.
- **high volume.**
- low pitch

11. When you increase the volume of a sound, the sound gets ________.

1/2

- **louder**
- quieter

12. When you decrease the volume of a sound, the sound gets ________.

1/2

- louder
- **quieter**

Level 3: Application: I can apply my knowledge of sound to explain why sounds get louder or quieter as the distance from the source changes.

Required: 3/3 **Pupil Navigation:** on
Randomised: off

13. When you stand close to the source of a sound, the volume of the sound is higher than when you stand far away, so the sound is ________.

1/2

- louder
- quieter

14. When you stand far away from the source of a sound, the volume of the sound is lower than when you stand close to it, so the sound is ________.

1/2

- louder.
- quieter.

15. When you stop the source of a sound from vibrating, the sound _______.

1/3

- stops
- gets quieter
- gets louder

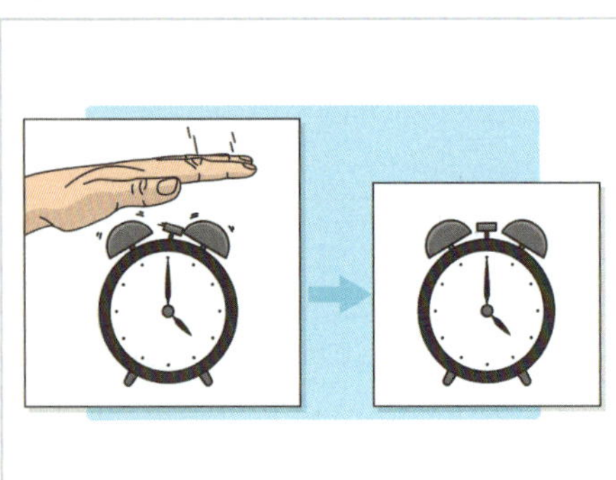

Level 4: Analysis: I can use data to explain real-life scenarios.

Required: 3/4 **Pupil Navigation:** on
Randomised: off

16. Hannah puts a siren on the floor. She uses a data logger with a sound sensor to measure the volume of the sound as she moves away from the siren. What does Hannah's graph show happening to the volume of the sound as she moves away from the sound source?

1/3

- **The volume of the sound increases as Hannah moves away from the source.**
- **The volume of the sound stays the same as Hannah moves away from the source.**
- **The volume of the sound decreases as Hannah moves away from the source.**

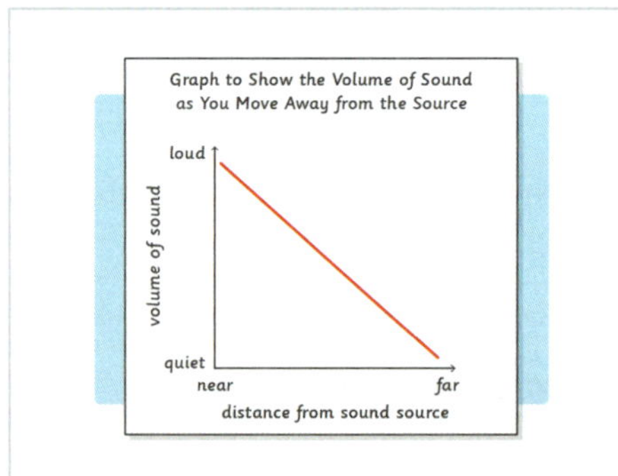

17. Miss Shah leaves a data logger with a sound sensor in the classroom to record the volume of sound in the room during the school day. The graph of her results shows that at 8 am, when the classroom was empty, the volume of sound was 30 decibels (dB). Use the graph to suggest another **two** times that the classroom may have been empty.

2/6

- 9 am
- 11 am
- 12 pm
- 1 pm
- 3 pm
- 4 pm

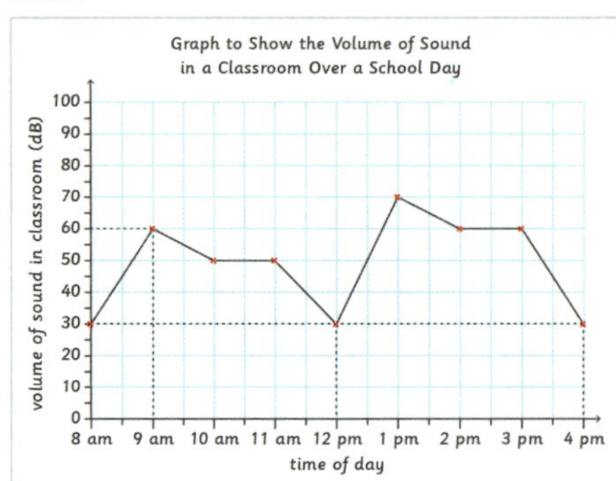

Level 4 *continued*

18. ☐ ☒ ☐ 1/4 Miss Shah leaves a data logger with a sound sensor in the classroom to record the volume of sound in the room during the school day. The graph of her results shows that the sound in the classroom was loudest at 1 pm. Use the graph to find the volume of sound in the classroom at 1 pm.

- 30 dB
- 50 dB
- **70 dB**
- 90 dB

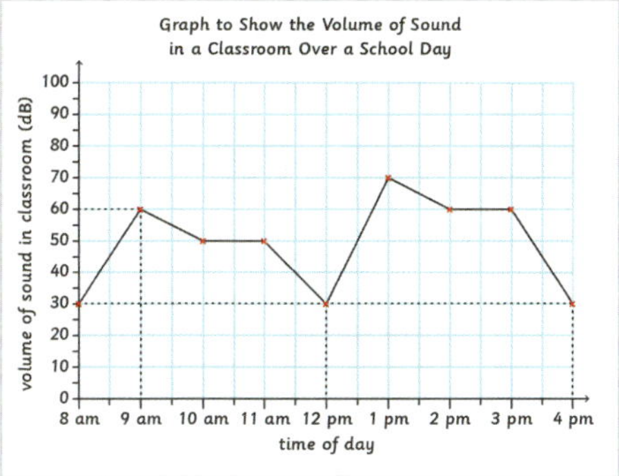

19. a b c These photographs show 'The Singing Ringing Tree' sculpture. It stands on a windy hillside in Lancashire and is made from pipes of different lengths, arranged in different ways. Sometimes the sound from the sculpture can hardly be heard, but at other times it howls loudly. Use your knowledge of sound to explain how the wind affects the volume of the sound made by the sculpture.

- Open question, no set answer

The Singing Ringing Tree, Lancashire

Singing Ringing Tree by Wendy Mann; *ringing tree* by Henry Brett

Sound

Objective: I know how sounds are made and how they travel.

Quick Search Ref: 11251

Level 1: Knowledge: I can describe how a sound is made.

Required: 6/6 **Pupil Navigation:** on **Randomised:** off

1. We hear a sound when vibrations from a vibrating object enter our __________.

1/3

- **ears**
- eyes
- nose

Sound Waves Travelling

2. A sound is made when an object __________.

1/3

- **vibrates**
- moves
- jumps

A Gong Vibrating to Make a Sound

3. A vibration is a repeating backwards and forwards __________.

1/2

- sound
- **movement**

A Gong Vibrating to Make a Sound

4. When you put your hand on your throat while you are talking or singing, you can feel the __________ that make the __________ of your __________.
Sort the words into the correct order to complete the sentence.

- **vibrations**
- **sound**
- **voice**

Vibrations Made by the Voice Box

5. A __________ is made when an object vibrates. The vibrations __________ as sound __________.
Sort the words into the correct order to complete the sentence.

- **sound**
- **travel**
- **waves**

Sound Waves Traveling

Level 1 *continued*

6. Leah is struggling with her homework on sound. Sort the words into the correct order to complete her sentences.

- ears
- vibrates
- movement
- waves

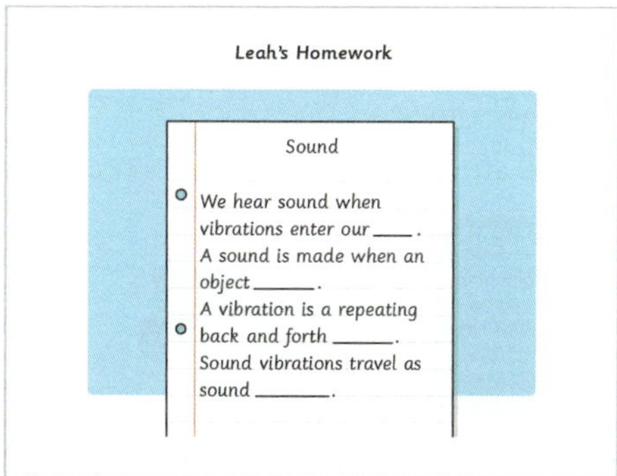

Level 2: Understanding: I understand how sound travels.

Required: 7/7 **Pupil Navigation:** on

Randomised: off

7. Sound __________ have to __________ through a substance. The substance can be solid, liquid or __________.

Sort the words into the correct order to complete the statement.

- waves
- travel
- gas

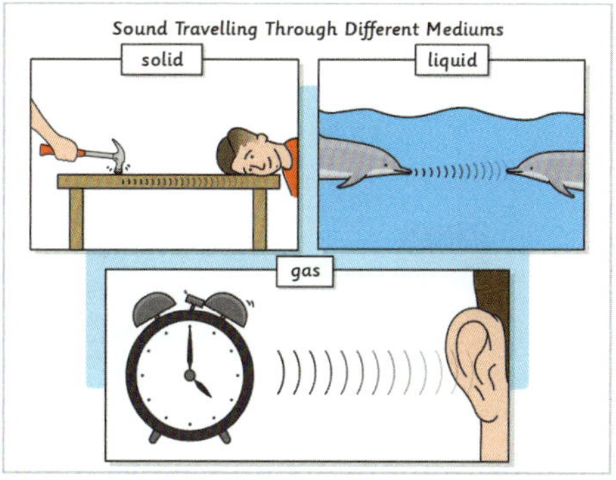

8. Can sound travel through space?

Select yes or no and a reason for your answer.

2/4

- no
- yes
- because space is full of gas that can vibrate
- because there is no air in space to vibrate

9. Sound __________ are vibrations that travel from a sound source.

- waves

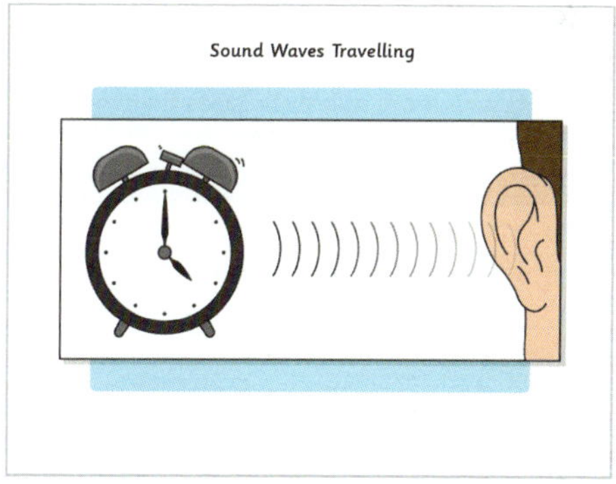

10. Sound waves can bounce off solid objects. Reflected sound waves can be heard as an __________.

- echo

Level 2 *continued*

11. The speed of light in air is 300,000,000 metres per second. Which travels faster in air: light or sound?

1/3

- **light**
- **sound**
- **Light and sound travel at the same speed.**

Speed of Sound Through Different Substances

substance	speed (m/s)
air	343
water	1,483
steel	5,940

12. Sound travels at different speeds through different substances. Which of these statements does the table suggest is correct?

1/3

- **Sound travels at the same speed through all types of substances.**
- **Sound travels more quickly in gases than in liquids.**
- **Sound travels more quickly in solids than in liquids or gases.**

Speed of Sound Through Different Substances

substance	speed (m/s)
air	343
water	1,483
steel	5,940

13. Which **three** of these substances can sound travel through?

3/4

- **air**
- **water**
- **glass**
- **space**

Level 3: Application: I can apply my knowledge and understanding of sound to real life situations.

Required: 3/5 **Pupil Navigation:** on

Randomised: off

14. Mica has attached a long plastic ruler to the desk. He pushes the ruler down and lets go. Why does Mica hear a sound when he lets go of the ruler?

Sort the statements into the correct order.

- **The ruler vibrates.**
- **This vibrates the air around the ruler.**
- **The vibrations travel as a sound wave.**
- **The sound waves enter Mica's ear.**

15. Luca is listening to the sound from his radio. The sound waves make Luca's eardrums ____________.

- **vibrate**

Level 3 *continued*

16. Ethan's brother is banging a drum loudly. Describe how the loudness of the sound that Ethan hears depends on the distance between him and the drum.
Use the illustration to help you.

- Open question, no set answer

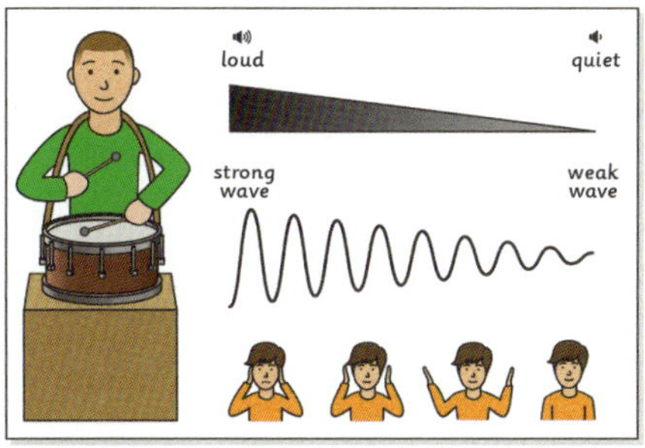

17. Ethan's brother is banging his drum loudly. Explain why the sound that Ethan hears is louder when Ethan is closer to the drum.
Look at the sound wave on the illustration to help you.

- Open question, no set answer

18. Hummingbirds are very small birds that get their name from the way that they flap their wings very quickly. Where might the name 'hummingbird' come from?

- Open question, no set answer

Hummingbird

hummingbird by Roger Ahlbrand

Primary Science Y3/4

Electricity

Circuits

Circuits Vocabulary

Objective: I can understand vocabulary related to electrical circuits.

Quick Search Ref: 11034

Level 1: Knowledge of circuits vocabulary: I can spell and define scientific words for circuits.

Required: 10/10 **Pupil Navigation:** on **Randomised:** off

1. Microwave ovens, TVs and hairdryers are all electrical ________.

1/3

- conductors
- **appliances**
- circuits

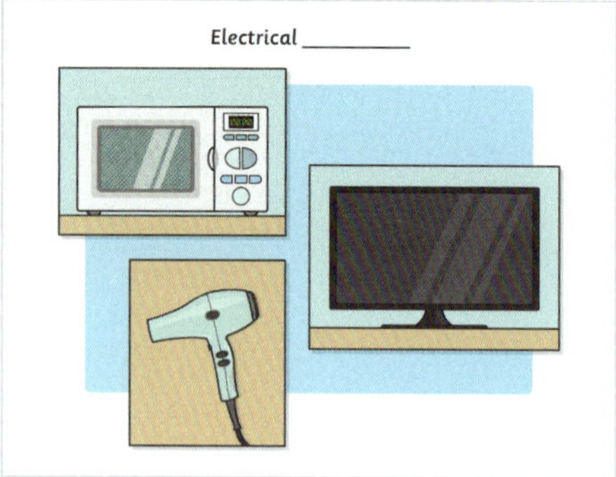

2. An electrical circuit is . . .

1/2

- **cells, wires and other components joined together in a loop.**
- made up of only wires.

3. A cell is a source of ______ in a circuit.

- energie
- **energy**
- power
- electricity
- enagy

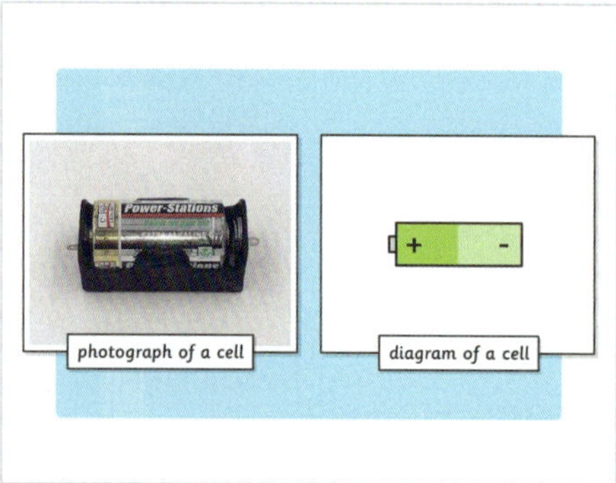

4. Wires . . .

1/3

- light up in a complete circuit.
- **connect a cell and other components together in an electrical circuit.**
- are a source of energy in an electrical circuit.

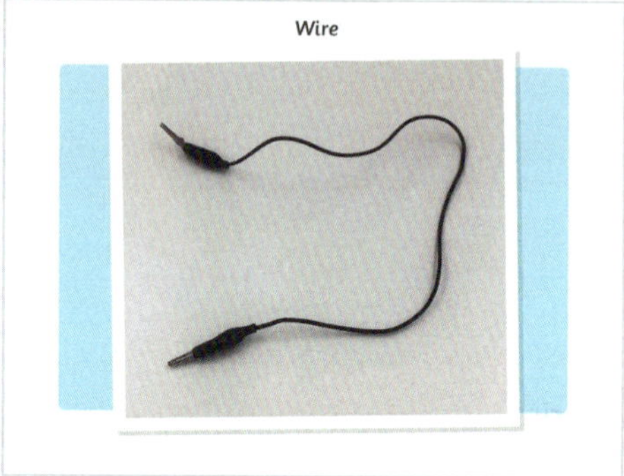

5. A buzzer connected in a complete circuit . . .

1/3

- spins.
- lights up.
- **makes a sound.**

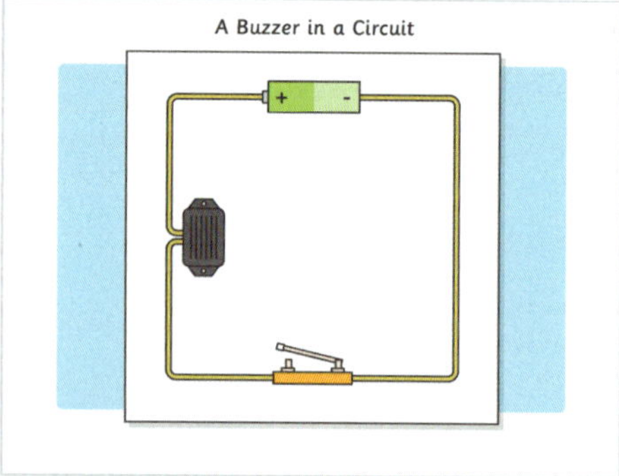

6. A ______ lights up when it is part of a complete circuit.

- **lamp**
- **bulb**
- **light bulb**

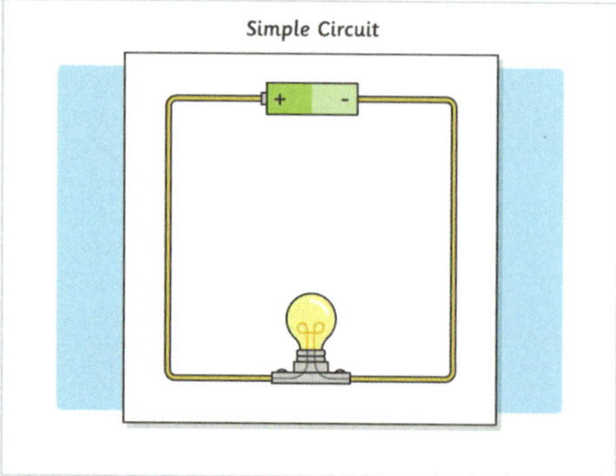

Level 1 *continued*

7. Electricity that we access in our homes through electrical sockets is called ______ electricity.

a b c

- mains

8. An object that electricity can flow through is called an electrical conductor. An object that electricity can't flow through is called an electrical _ _ _ _ _ _ _ _ _.

a b c

- insulator.
- insulator

9. A _______ can be open or closed. When it is closed, it completes the circuit and allows electricity to flow around the circuit.

1/3

- lamp
- switch
- cell

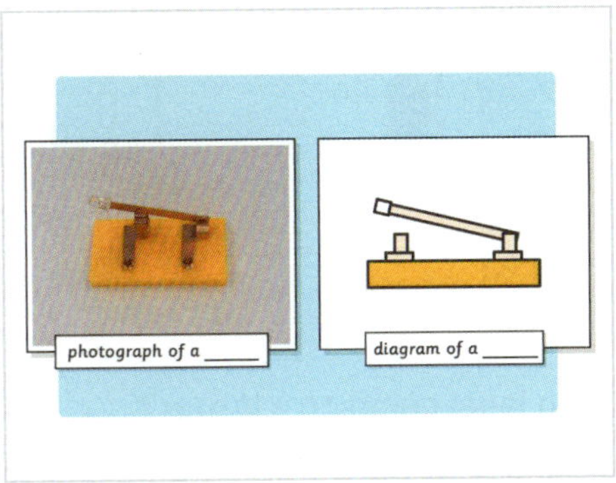

10. _ _ _ _ _ _ _ _ _ _ _ is energy that is carried around a circuit.

a b c

- Electricity
- Electrisity

Electrical Circuits

Objective: I can identify and name the basic parts of a circuit, including cells, wires, bulbs, switches and buzzers. I can identify whether or not a lamp will light in a simple series circuit, based on whether or not the lamp is part of a complete loop with a battery.

Quick Search Ref: 10893

Level 1: Knowledge: I can identify and name the components in a series circuit.

Required: 8/10 | Pupil Navigation: on | Randomised: off

1. An electrical circuit is . . .

1/3

- made up of only wires.
- how electricity is made.
- **a cell, wires and other components joined together in a loop.**

2. An electrical circuit is made up of wires, a cell and other __________ joined in a loop.

1/3

- **components**
- wires
- parts

3. A battery acts as a source of ________ in an electrical circuit.

- electricity
- **energy**
- power

4. Which label points to a cell?

1/5

- **A**
- B
- C
- D
- E

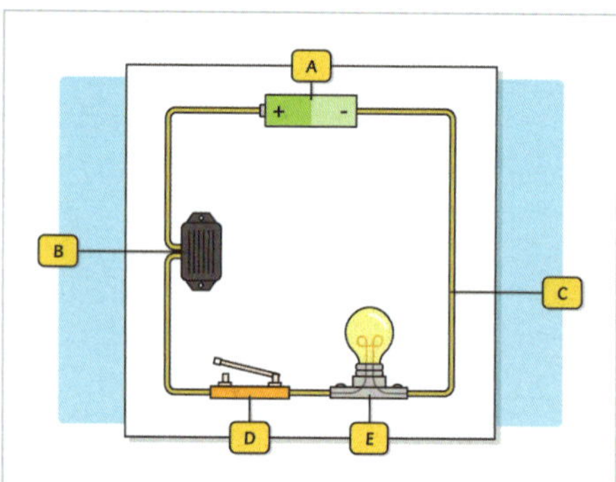

5. Label B points to a _ _ _ _.

- wyre
- whire
- switch
- **wire**
- wir
- lamp
- cell
- wiyre

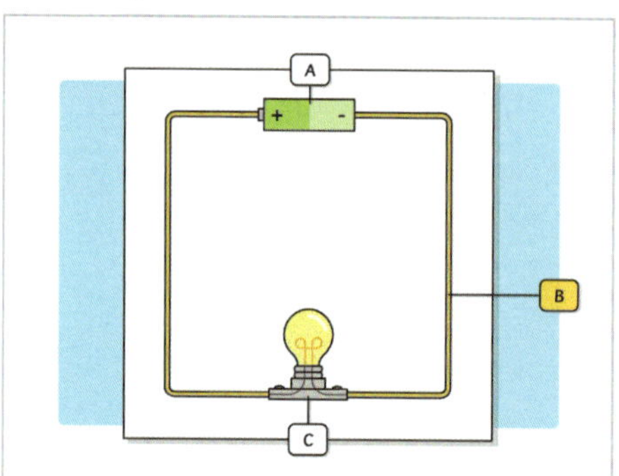

6. Which label points to a lamp?

1/4

- A
- B
- C
- **D**

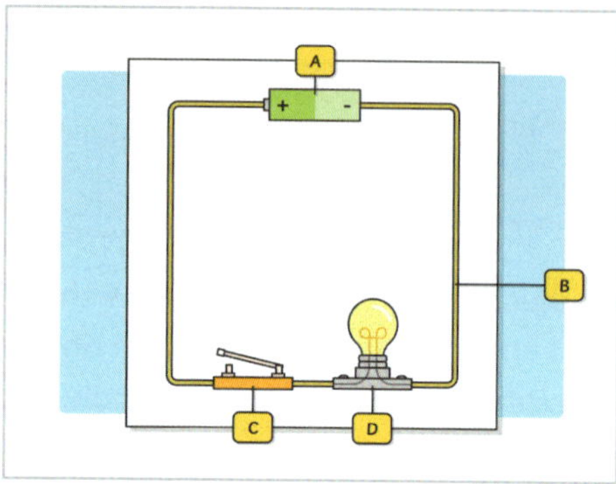

7. Which component of the circuit does the label point to?

1/5

- buzzer
- **switch**
- wire
- cell
- lamp

8. Which label points to a buzzer?

1/4

- A
- B
- **C**
- D

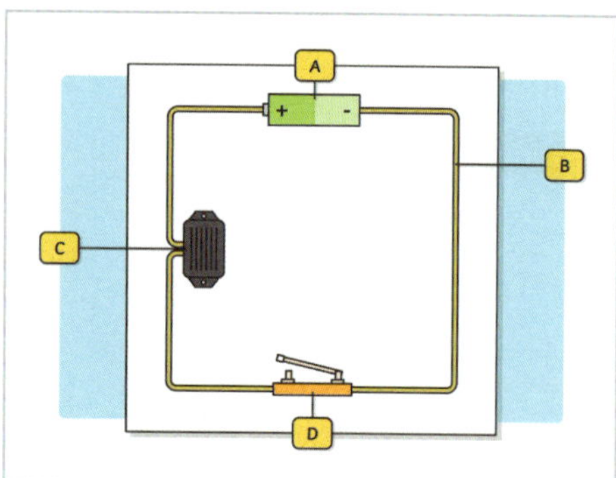

Level 1 *continued*

9. What is an electrical circuit?

1/3

- wires connected together
- **a cell, wires and other components joined together in a loop**
- how electricity is generated

10. What is a component?

1/3

- a wire in an electrical circuit
- the source of energy in a circuit
- **a part of an electrical circuit**

Level 2: Understanding: I can explain the role of each component in a series circuit.

Required: 6/6 **Pupil Navigation:** on
Randomised: off

11. A cell is a source of ______ in a circuit.

abc

- electricity
- **energy**
- power

12. A battery is made up of two or more . . .

1/3

- **cells.**
- switches.
- wires.

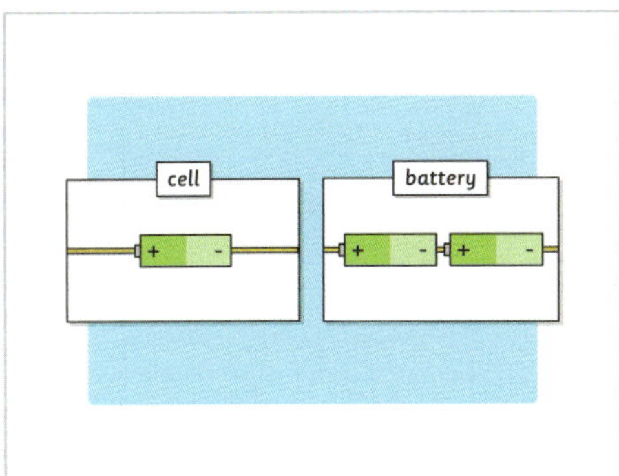

13. A wire . . .

1/3

- lights up in a complete circuit.
- is a source of energy in an electrical circuit.
- **carries electricity between components in an electrical circuit.**

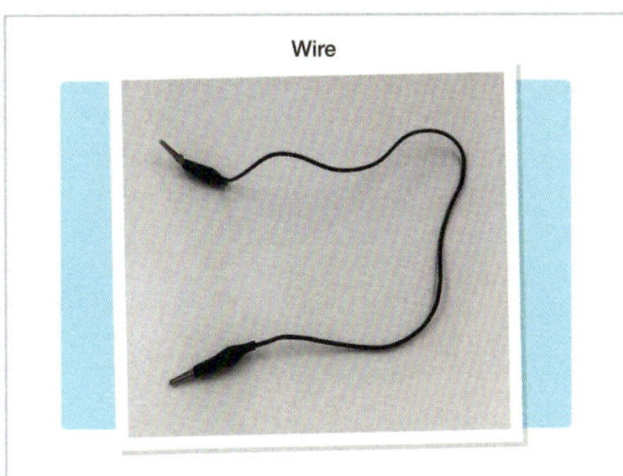

14. What component lights up when it's part of a complete circuit?

abc

- **bulb**
- **a bulb**
- **a lamp**
- **lamp**
- **the lamp**
- **the bulb**

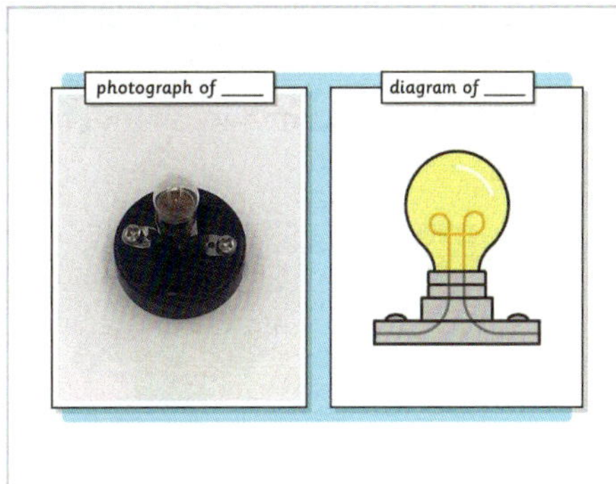

15. A switch can be open or closed. When it's ________, it completes the circuit and allows electricity to flow around the circuit.

1/2

- open
- **closed**

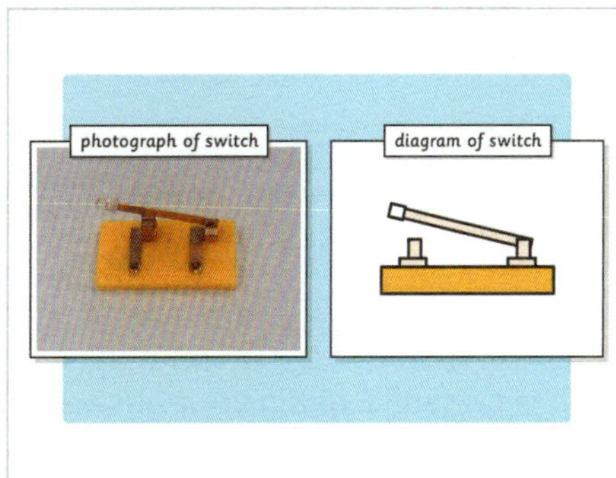

16. A buzzer _______ when connected in a complete circuit.

1/3

- spins
- lights up
- **makes a sound**

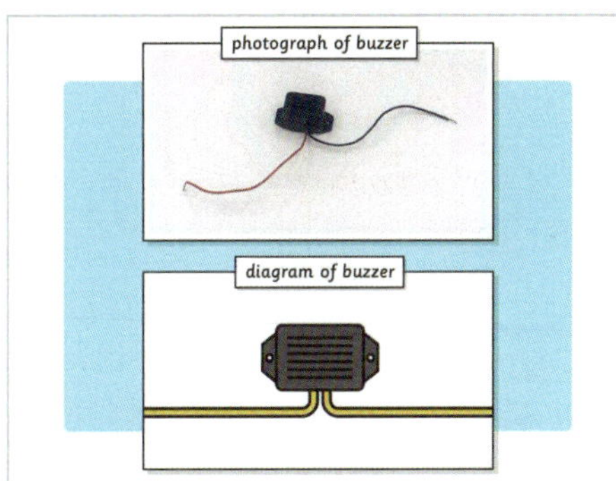

Level 3: Application: I can apply my knowledge and understanding of electrical circuits to the world around me.

Required: 3/4 **Pupil Navigation:** off
Randomised: on

17. Charlie, Jade and Ryder are talking about how safe it is to use a cell in an electrical circuit. Who is right?

1/3

- Charlie
- **Jade**
- Ryder

18. Which of these everyday items need electrical circuits to work?

4/6

- **tablet computer**
- glow stick
- **remote control**
- **digital watch**
- bicycle
- **ceiling light**

19. Will the lamp in this circuit light up when the switch is closed? Explain your answer.

- Open question, no set answer

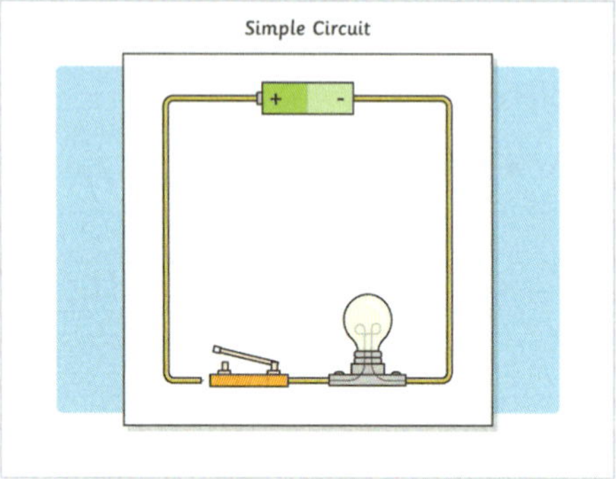

20. Jade's circuit won't light up. What could be the reason for this?

1/4

- The lamp has blown.
- The cell is flat.
- One of the wires is broken.
- **All of these are reasons.**

Specifications

Charging & Storage Cabinet

- 33-bay up to 10” tablet charging cabinet
- 2 easy access sliding shelves
- 4 efficient fans for ventilation
- locking doors with keys
- 4 castors / 2 handling bars overload, leakage and lightning surge protection
- unladen weight 96kg
- CE / ROHS / FCC compliancy
- 3 years warranty

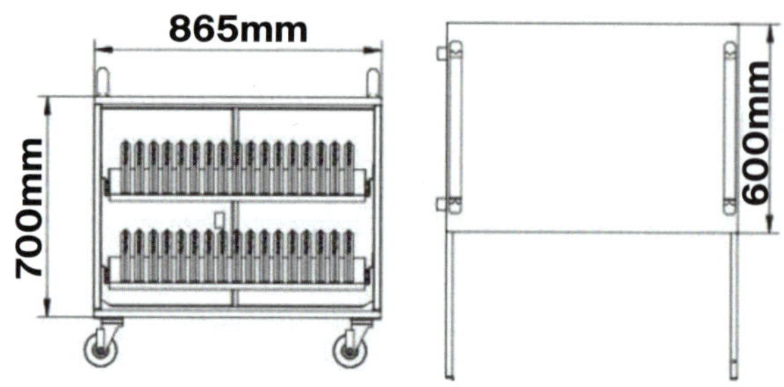

Student and Teacher Tablets Configuration

All tablets with LbQ Tasks app pre-loaded and tablets installed in cabinet including charging cables for quick and easy deployment in classrooms.

	8” Android Tablet with Protective Cover	10” Android Tablet with Protective Cover
Display		
Resolution	1280 x 800	1920 x 1200
16:10 display ratio	✓	✓
Capacitive 5-touch	Capacitive 5-touch	Capacitive 10-touch
System		
Cortex 64bit Quad Core 1.5GHz CPU	✓	✓
2GB of RAM	✓	✓
16Gb of storage	✓	✓
Android 7.0	✓	✓
Front and rear Camera	✓	✓
Input / Output Ports		
1 x Micro SD Slot	✓	✓
1 x Micro USB (PC / device / charger)	✓	✓
Micro-HDMI output	✓	✓
1 x Earphone, 1 x Speaker, 1 x Mic	1 x Earphone, 1 x Speaker, 1 x mic	1 x Earphone, 2 x Speaker, 1 x mic
Communication		
Wifi - 802.11a/b/g/n	2.4 GHz / 5 GHz	2.4 GHz
GPS module	✓	✓
Bluetooth	✓	✓
Power		
5V 2A	✓	✓
Battery	3500 mAh battery	7000 mAh battery
Physical		
Colour: metal black	✓	✓
Weight	300g	560g
Dimensions	205 x 114 x 8mm (approx.)	263 x 164 x 9mm (approx.)
Warranty		
3 years advanced replacement for faulty tablets - does not cover damage CE / ROHS / FCC compliant	✓	✓

Learning by Questions